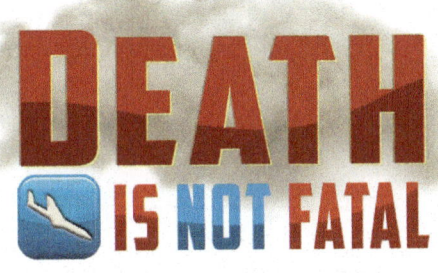

Vernon L. Grose, D.Sc.

An exhaustive review and analysis of the meaning of death based on the author's expertise in the systems approach to managing risk coupled with his international expertise as an authority in directing investigation of aviation and space disasters

Art Design and Cover Illustration: Bradley W. Grose

Copyright © 2016 by Vernon L. Grose, DSc

DEATH Is Not FATAL
by Vernon L. Grose, DSc

Printed in the United States of America.

ISBN 9781498472234

All rights reserved solely by the author. The author guarantees all contents are original and do not infringe upon the legal rights of any other person or work. No part of this book may be reproduced in any form without the permission of the author. The views expressed in this book are not necessarily those of the publisher.

Unless otherwise indicated, Scripture quotations taken from the Holy Bible, New International Version (NIV). Copyright © 1973, 1978, 1984, 2011 by Biblica, Inc.™. Used by permission. All rights reserved.

www.xulonpress.com

Dedication

To Phyllis, my incomparable wife for 65 years
who has selflessly stood by me in support
of the diverse activities and interests
that made this book possible

Contents

FOREWORD..ix

Chapter 1. Human Death as a SYSTEM 13
Chapter 2. The Meaning of Risk 28
Chapter 3. Human Obsession With Death 46
Chapter 4. Death: The Wrath of God? 65
Chapter 5. Demise of Noble Savage Theory 72
Chapter 6. Modern Masking of Death 80
Chapter 7. Death: Finality or Change of State? 95
Chapter 8. One Answer From Antiquity 109
Chapter 9. Political Solutions for Death 126
Chapter 10. Toward a Rational Forum....................140
Chapter 11. Death in the 21st Century 158
Chapter 12. Managing Your Risk of Death 178
Appendix A – LIFE'S FOUR SEASONS 199
Appendix B – Everybody Dies 203
Appendix C – Bibliography of Referenced Books...........205
Index...207

FOREWORD
DEATH Is Not FATAL

Benjamin L. Aaron, M.D.

Death is a concern of all of us and hovers about us like a pesky fly that won't go away. As an abstract idea, we can deal with it in ways that mute the ultimate reality of it, always keeping it at arm's length.

There are many contradictions in our attitude toward death. For example, we all know that we must die, but which of us would be willing to name the day and hour in advance, even if we could. We want to live a "long time," but no one wants to live forever (on this earth). We know that death is necessary for population control on this small planet, but which of us would volunteer to make the numbers fit? Dr. Grose considers death in its many aspects in this text and provides for the reader a perspective and a framework to come to terms with death in his or her particular world view through considering death as a system.

Physicians are confronted daily by the need to manage the risk of death, often a very delicate and difficult task. Recognizing that unduly prolonging life is not always the kindest or most humane outcome, and having the management tools (ventilators, antibiotics, pacemakers, etc.) to do so, places the physician in the unenviable role of making life and death decisions.

As a physician and surgeon, I have many times had the opportunity and responsibility of managing the risk of death and have employed the risk management principles espoused by Dr. Grose. My knowledge and expertise in this effort came by the laborious and inefficient "on the job training" mode, and

I think that the information and methodology herein contained would have been of great assistance in my learning years.

A case in point was the celebrated attempted assassination of President Ronald Reagan in March of 1981, in which I became the manager of his precarious health for eleven days. Clearly, the intent of the assailant was that the President should die, and he set the stage for this very effectively as he delivered a bullet into the President's left chest which narrowly missed his heart and lodged in his lung, producing massive bleeding. Here then, with the President of the United States at risk for dying, is a challenge of risk management with colossal stakes, as a surgeon pits his skill at healing against the formidable skill of the assailant to cut short a life.

The President had collapsed on arrival to the emergency room, indicating a loss of at least one third of his blood volume. With blood continuing to pour forth from his chest, the options available were: (1) wait him out, as bleeding from the lung often slows or stops, making up the loss with blood transfusions, (2) emergently operate with direct control of the bleeding.

Option One had the obvious advantage of avoiding the risks of anesthesia and surgery in a 70-year-old man who was already treading heavily on his margin of survivability.

Option Two, though having more risk (potential for loss) in the short term, provided for controlled termination of the bleeding.

The management principle I applied was that the known risk of an operation with the surety that the bleeding would be stopped provided a more predictable route of recovery. A wait-and-see approach (Option One), with the imponderables of additional critical blood loss, undrained blood remaining in the chest, the ensuing possibility of poor lung function and infection, as well as leaving a foreign body (the bullet) in the President's lung, seemed a poor choice by comparison.

As it turned out, the later revelation that the bullet contained a toxic substance, called lead azide, would have mandated a return to the operating room to remove it, had I not moved ahead with the early surgery. Fortune thus favors, not by chance, good risk management.

The oft-used statement, "time is risk," (i.e., provides opportunity for loss), applies especially to the operating room, where unnatural forces are being utilized to provide a safe passage for a patient on a hazardous journey. Thus, having controlled the bleeding, when the tiny, flattened bullet was difficult to find in the spongy lung, I came up against another pair of options: (1) leaving the bullet and terminating the surgery quickly, or (2) spending the additional time to assure its location and removal.

Things were going well, so it seemed prudent to elect Option Two and spend the time necessary to remove the bullet which could be a source of further bleeding and infection, as well as recover the single bit of tangible evidence that tied the would-be assassin to the victim.

The process of risk/benefit tradeoff, it appears, can involve many relevant elements, depending on the particular circumstances — requiring the manager to assess and weigh the elements, negative and positive, that affect a satisfactory outcome.

Recovering from the injury and the surgery, the President had fever and internal bleeding from his torn lung, raising the specter of infection on the one hand, and recurrence of the massive bleeding on the other. I then faced another set of options: (1) surgical re-exploration with resection of the injured portion of the lung, or (2) a conservative approach which would utilize intravenous antibiotics and airway clearing by way of bronchoscopy.

I considered this a very critical decision point, as the President was probably more vulnerable to problems (infection, bleeding or both) than at any time in his course. Both options had their obvious inherent risks, and had there been some problem with the function of the injured lung, I would have chosen Option One. As he was making progress otherwise, I carefully chose the conservative Option Two, electing to insert a bronchoscope into his airway and lung and remove the clotted, infected blood, thus clearing the small air passage and allowing the lung to ventilate more effectively. This was done at some discomfort (without anesthesia) and risk to the President, but on balance, in my judgment, was the best way to proceed, and his improved course reflected this wisdom.

There were a number of other critical, but less dramatic, decisions rendered in the President's case — each one involving a risk-versus-benefit tradeoff — considering all the relevant features of the particular situation. History has recorded the satisfactory recovery of the first President of the United States to survive the injury of attempted assassination.

As Dr. Grose so clearly illustrates, societal and individual attitudes toward death have changed over the centuries and will continue to change as we are affected by worldwide population blending, shorter time frames on all activities, a heightened pace of living, an aging population, encroaching technology, and enhanced expectations for good outcomes. In this environment, consideration of death sits patiently on our awareness and warrants education and enlightenment such as provided in this book if we are to effectively manage this ultimate reality of living.

"Get a grip" is prominent in today's lexicon of sophisticated communication. This book's treatment of the issue of death goes a long way toward enabling the modern thinker to get a grip on death that is applicable to the 21st century.

And paramount is Dr. Grose's contention that how you manage the first (spiritual) death — thereby determining where you spend eternity after the second (physical) death — may be the most important management decision of your whole life in this uncertain and insecure world.

Chapter 1

Human Death as a SYSTEM

"Death is not the worst that can happen to men." – Plato

"And what are *you* discussing tonight?"

Debi, one of the makeup cosmeticians in Washington's CNN studio, recognized me from many previous interviews.

The overhead TV monitor in the makeup room was announcing the latest news on the crash of ValuJet Flight 592 in the Florida Everglades. I had been rushed by limousine from my home to analyze the latest on-scene information.

I pointed at the monitor and replied, "The crash."

As CNN's Risk Analyst, I had already been on camera within hours of the disaster which had occurred the previous day. But Debi wasn't on duty then.

My answer drew a fast response from this attractive young mother of three: "I'm really *frightened*!"

"What's the matter?"

"Well, I just bought tickets to Cancun for my husband and me — the first time we've ever left our kids — on a *charter* airline. I wanted to surprise him. What do you think of charter airlines?"

"Well, their safety record certainly isn't as good as the major scheduled airlines. They're not subject to the same risk oversight."

She hadn't mentioned which charter airline she had booked, and I didn't really want to scare her. But since she had asked, I felt obliged to at least share my professional opinion with her.

"Oh, that makes me even *more* concerned! Maybe I should buy tickets for two separate flights so that at least one of us would survive for our kids."

Just as Debi finished my makeup, Jeanne Meserve arrived for Debi to apply makeup on her. The anchor person for CNN's "*World News Tonight,*" Jeanne would be interviewing me shortly.

So, while Debi had Jeanne in the chair, I used the opportunity to rapidly list and review factors about the crash that the two of us would be discussing on camera in a couple minutes. Of course, Debi couldn't help but listen to those crash details and failures as she kept applying Jeanne's makeup.

As Jeanne then left in a hurry to go on the set, I stayed behind just a moment to offer Debi some solace — since she was obviously experiencing great distress.

"Debi, although I certainly understand and empathize with your anxiety about flying, being anxious really isn't going to help you. You need to be ready to *die!*"

Her reaction was immediate and startling: "I *know* it — and I'm *not* ready!"

Debi is like most of us. There are so many factors to consider about death — derived from folklore, family history, attorneys, terrorism threats, wives tales, rumors, news media, horror stories, and financial planners.

While she knew that she wasn't ready to die, how would she *get* ready? What would she have to *do*? Whom would she have to *consult* — if anyone? When could she be assured of being *fully ready* to die?

Debi's Dilemma

Believe it or not, my concern for Debi became a strong stimulus for me to consider applying my professional expertise — systems methodology — to the dilemma of human death.

That very brief encounter with Debi continued to provoke me. So much so that it became a major reason for writing this book. I also wanted to share with many others that it is possible to become ready to die.

Beyond that, I was convinced that obtaining a solution to the "risk of death" has great influence on a person's entire life. Debi's predicament wasn't hers alone. *Everyone* must resolve that quandary for themselves. So if I could help others in their search for being "ready to die," I wanted to offer it.

Yet that wasn't the way I personally became ready to die. And I knew that a cold, rational, impersonal method to sort out all the assets and debits of life could never solve Debi's dilemma either.

So my desire to recount and share my resolution of this perplexing matter continued to mount after that short encounter with Debi. It occurred to me that perhaps my solution could become yours as well.

Every one of us obviously must work out our own unique aspects of "getting ready to die" — whether we are an atheist, agnostic, or a believer of some sort. In our culture, matters like facing death are very private. But, on the other hand, we are always learning from others, too. Could it be that my answer to that dilemma would help anyone else?

I wrestled with the idea. Finally it seemed that taking the gamble of exposing failures, errors, and defeats in my own struggle to reach assured readiness about my own death would be worthwhile only if readers were able to apply those "lessons learned" in their personal quest for the same type of certainty.

Pervasive Death

It has only recently occurred to me that most of my professional life has involved death. No, I'm neither a physician, executioner, nor mortician. But being an expert in analyzing and managing risk means that much of my work is measured against the *ultimate* risk — death.

In retrospect, even a risk management technique that I developed and have implemented in a wide range of business, industrial, and government settings for many years often weighs death as the ultimate risk. It is known as **SMART™ (*Systems Methodology Applied to Risk Termination*).** However, that doesn't always mean *human* death. It can also mean death of an *investment, project, reputation, product,* or *market*.

For many years I've employed SMART™ to provide answers in numerous diverse situations involving risk that seemed to defy rational solution. For example, it was applied successfully to control terrorism at the 1984 Olympics in Los Angeles – following the terrorist massacre in Munich during the 1972 Summer Olympics when the Palestinian group *Black September* took 11 Israeli Olympic team members hostage and later killed them. It was also employed for managing common corridor risk in the design of the Washington, DC METRO system as well as nuclear power risk following the 1979 Three Mile Island nuclear accident.

Earlier Approaches to Death

Professionally, the ultimate risk of death has always been an underlying issue for me. But it is seldom addressed directly. For example, I was aware of Thanatology — the scientific study of death that investigates the mechanisms and forensic aspects of death, such as bodily changes that accompany death and the post-mortem period, as well as wider psychological and social aspects related to death.

And there are a host of books on *preparing* for death as inevitability. On that aspect, the work of Swiss psychiatrist Elisabeth Kubler-Ross and her 1969 book *On Death and Dying – What the Dying Have to Teach Doctors, Nurses, Clergy and Their Families*[1] is legendary.

Yet all these studies, pursuits, schemes and books — admirable as they are – lacked the specific answer Debi needed to be ready to die. Her sudden, agonizing declaration continued to provoke me.

While I have never seen any attempt to do so, I believe it is not only possible but imperative to address human death *systematically*! So you are invited to share my high confidence that the systems approach will prove helpful in managing human death – even yours.

Today, the word "system" is so overused that it has almost lost all meaning or significance. Anything connected with technology, computers, and popular hand-held devices are somehow linked with the word "system." Even losing weight is linked to and advertised as a "system." So are criminal justice, interstate highways, political elections, and public health.

However, the systems approach is actually quite ancient – having hidden incognito for centuries.

The Systems Approach: Old Idea, New Name

Over 2,000 years ago, Greek thinkers conceived the idea of viewing any complex subject, setting, event, or undertaking as a *bounded whole*. Though it may have consisted of many diverse parts or elements, they still considered it as a *singularity* — surrounded by recognized boundaries.

Let's just call it "the Box." On one side, the Box had *known inputs* — conditions, values, resources, events, forces, and factors that either establish or influence everything in the Box. On the other side, the Box had *desired outputs*, the intended results of whatever activity took place inside the Box.

By using this technique, these ancient philosophers could "get their arms around" any complicated matter and — by that simplicity — overcome the otherwise overwhelming impossibility of understanding and mastering it. That process has even been described as "womb-to-tomb" thinking whereby everything relevant to a subject is embraced, examined, and evaluated.

Over many centuries, this pattern of thought seemed to fall from use and recognition. However, during the mid-20th century when technology began to proliferate and make life increasingly complicated, the archaic concept was recalled, dusted off, and renamed "the systems approach."

Perhaps the first and most recognizable sign of its resurrection occurred during World War II when the United States was fiercely competing with the Axis nations to produce the first atomic bomb.

The Manhattan Project was assigned that job — and used an early version of what was to become the systems approach. Why? Because there was a critical need to cut across all governmental and industrial bureaucratic fiefdoms to assemble the right people to focus on a single objective – a bomb that had never existed.

It wasn't very democratic. The key to success lay in the ability to identify, obtain, and organize the diverse specialists into a team that could develop that fearsome weapon in minimum time. Efforts and specialties of academicians, government bureaucrats, military and political leaders were all amalgamated and subordinated to a single objective – foregoing their primary individual interests.

That centralized, focused approach, with its sharply-defined objective and autocratic authority, proved to be essential to winning the race for atomic dominance and thereby quickly ending the war.

But that approach failed to be adopted by business and industry after World War II. As technology proliferated, intolerable management fragmentation was produced by industrial and governmental specialization that increasingly precluded efficient, effective completion of large projects.

Specialists of all kinds in the early 1950's had lost sight of the "big picture" to which they were only contributors. Executives were forced to treat those contributions as stand-alone "silos" or "stovepipes" that defied subordination to a greater whole.

Consider a military aircraft as an example. It consists of an airframe, engines, specialized electronics, payload, guidance, and crew accommodation. However, it defied being managed as a single entity. Why? Because all those contributing elements or silos had their own specialized values, terminology, weight, cost, production time, and requirements. They focused all their effort on their *specialty* rather than on the ultimate *aircraft* that utilized that specialty.

Though those activities all generated risks unique to their specialty, there was no way to integrate and manage the *collective impact* of those particular risks on the performance of the larger entity in which the specialty was incorporated. As shown in Figure 1, the executive responsible for the entire aircraft was faced – due to independent silos — with the impossibility of balancing and managing many diverse risks. Worse yet, all those various risks represented very disparate viewpoints about risk!

Impact of SILOS
(Functional Specialties)

- Diverse **definitions** of risk
- Unbalanced **emphasis** on perceived risk
- **Disconnection** from corporate risk strategy
- Failure to comprehend risk **interdependence**
- Counterproductive **inefficiency** of risk control

Figure 1- *Impact of SILOS*

Further, there was no known means for integrating all those diverse specialties to accomplish a singular, focused mission. It was this disturbing situation into which the systems approach was born – resurrecting and implementing a concept over 2,000 years old!

Complexity's Fragmentation

The effectiveness of traditional schemes of management that had been beneficial earlier in the century had proven inadequate. Why? Because there was increasing proliferation of "a government bureau for every specialty." That was the same problem that had forced the creation of the Manhattan Project for the atomic bomb.

It seemed that every time a new specialized endeavor arose — whether in science, engineering, banking, medicine, insurance, or transportation, it was immediately "organized" into an entity with a title, organizational structure, advocates, heroes, annual meetings, publications, awards, and unique identity. This pattern was apparently the only means for acknowledging and accommodating the complexity that inevitably accompanied progress.

So progress produced complexity. And complexity produced specialization. Experts began to know more and more about less and less. This explosion of specialized endeavors — while enabling knowledge to expand at an every-increasing rate — defied the possibility of an *integrated focus of specialties on a common task*.

The day of the Renaissance Man faded as the ability to grasp wholeness disappeared. That need for wholeness then set the stage for the re-discovery of the old Greek concept but with a new name — the *systems approach*.

Another way to view specialization is to see it as a *fragmenting force* that destroys unity or holism. It must be harnessed or counter-balanced by generalization before it can make its contribution.

Those *specialists* who "know more and more about less and less" must be matched with *generalists* who deliberately determine to "know less and less about more and more" so as to rise up, integrate those select specialties required to perform a major task, and then provide the leadership to realize its unified accomplishment.

Such unification is what any great symphony conductor performs. Even though that maestro may previously have risen to prominence as a renowned soloist, they set aside their specialty to unite all specialties to produce a masterpiece of sound.

By the mid-1950's, all US armed forces — starting with the US Air Force, followed by the US Navy and then the US Army — had adopted the systems approach. And its use was dictated by the need to manage weapon systems that had grown to be virtually unmanageable due their complexity, cost and competition.

Like the three US armed services, NASA had roots that were specialized. Its predecessor – National Advisory Committee for Aeronautics (NACA) — was formed in 1915 primarily for research. The result was a group of functional specialties that needed to be integrated and focused in terms of a mission. After NASA was formed in 1958, I was among those who early urged NASA to adopt the systems approach to manage all aspects of its manned space programs – particularly regarding risk.

Addressing a Government-Industry Conference at NASA Goddard Space Flight Center on the systems approach

However, NASA initially lagged in adopting the systems approach. It paid a significant price for failure to do so, too. On 27 January 1967, fire in Apollo 1 capsule being tested on the pad killed the 3 astronauts inside. This tragedy provoked adoption of the systems approach for re-design of the spacecraft.

Two radical changes resulted. First, the stowed inward-opening crew hatch became fixed and outward-opening. Second, the spacecraft cabin pure oxygen environment was replaced with a two-gas system that took 14 months of redesign costing over $100 million and added 1 ton in weight.

As the Apollo missions to the Moon grew in complexity, the importance of systematic management increased. Dr. Wernher von Braun, Director of NASA's Marshall Space Flight Center, appointed me in 1969 to the 5-man NASA Safety Advisory Group for Space Flight to "ensure absolute safety of our manned flight missions."

Discussing aspects of NASA's Space Shuttle with Dr. Wernher von Braun on 30 January 1976

There is little doubt that the systems approach enabled the remarkable success achieved in landing men on the Moon and returning them safely to Earth.

The Secret of Centrality

The systems approach itself — though apparently so simple and uncomplicated — has been harried by ambiguity since its reappearance on the scene. In the nearly six decades since the ancient Greek idea was renamed, it has yet to be universally *defined*. Its characteristics or common properties are quite widely conceded and professed. But it awaits an undisputed definition.

However, we can still employ the concept. So let's start with its *properties*:

o   A "reason to be" or *intended objective*

o   A *singularity* or wholeness built from diversity

o   An *interaction* or activity among many parts

o   A *beginning-to-end* process or "throughput"

Those properties collectively describe what we are going to describe and apply as "the systems approach" to human death. Understandably, you may find this goal to be something a bit challenging. But let's approach the idea together.

NASA – System *Cause Celebre*

Perhaps the greatest demonstration of success for the systems approach ultimately occurred in the three manned NASA projects – Mercury, Gemini, and Apollo – that progressively resulted in attaining President Kennedy's famous national goal of "landing a man on the Moon and returning him safely to the Earth by the end of the 1960's."

After I had conducted two-week seminars on the systems approach at all eight NASA Centers, I was awarded NASA's "Silver Snoopy" medal by Apollo astronaut Tom Stafford.

USAF Brigadier General Tom Stafford – Commander of the Apollo-Soyuz 17-19 July 1975 rendezvous in space – presenting NASA's "Silver Snoopy" medal to me on 25 September 1974 in Houston

This medal was a unique honor because it can only be awarded by an astronaut. The three Soyuz Russian cosmonauts happened to be in Houston that day, preparing for the upcoming rendezvous mission. So I was introduced to them as well.

Creating a System

Debi's dilemma about death is actually quite similar to the dilemma faced by executives who were building aircraft before the systems approach was resurrected from ancient Greek thinking. Instead of defining an *aircraft* as a system, however, let us begin by defining *human death* as a system.

Figure 2 provides a universal definition for a system. Note that literally anything (from a coffee cup to the entire universe) can qualify to be considered a system.

> **SYSTEM**
>
> A **COMPOSITE** – *at any level of complexity* – of operational and support equipment, personnel, facilities, and software which are used together as an entity and capable of performing and / or supporting an operational role which results in changing **KNOWN INPUTS** into **DESIRED OUTPUTS**

Figure 2–*Definition of a System*

The foundation of the systems approach is both simple and profound. It consists of *bounding* or *circumscribing* whatever is intended to be addressed systematically.

The prescribed, confining limits must clearly remove any ambiguity about what *will* — and what *will not* — be considered a "system." As simple as it appears however, considerable skill and tenacity are required to unambiguously define a system.

Note that the "systems approach" begins with a clear definition and agreement on the limits, boundaries and description of whatever is declared to be a system. That process is an *art*, and few there are who initially understand its initiation and necessity. It requires mental discipline and does not immediately produce unanimity.

On the other hand, it is generally a pleasant experience – once comprehended and pursued to completion – for everyone who participates in realizing its existence. Further, it rewards the investment of time and thought with a readily-recognized and simple description of what would otherwise be fuzzy and vague.

Once the SYSTEM is defined and bounded, two major tasks become mandatory – establishing the system's KNOWN INPUTS and its DESIRED OUTPUTS.

Known Inputs

At the outset, it must be recognized that there are conditions, situations, forces or factors that collectively are responsible for bringing the system into existence. Had they not been present, the system would not exist.

These ambient phenomena are known as Known Inputs as defined in Figure 3. Together, they form the *argument* or *rationale* that explains how the system came to exist.

> **KNOWN INPUTS**
>
> Events, conditions, attitudes, needs, expectations, and situations that **ESTABLISH** the rationale for the system's existence.

Figure 3–Known Inputs

Just as *defining a system* is an art, so also is the *creation* of its known inputs.

Defining inputs requires a mindset that is not prevalent or readily obtained. Few have ever attempted to define inputs – primarily because *systems thinking* as a discipline is fairly rare. So patience and tenacity are involved. As a collective, Known Inputs comprise the "were it not for these" argument that lawyers would understand and use to defend a system's reality.

Experience has shown that limiting the number of inputs to no more than six or seven enforces the mental discipline to sharply refine and focus what otherwise might be a lengthy listing. This limit forces *truly major* facets to emerge – thereby providing executive decision-makers with readily-perceived understanding.

Desired Outputs

Desired Outputs in Figure 4 are almost the antitheses of Known Inputs – in the sense that they resolve, address or answer the establishment aspect of Inputs. Those outputs are always expressed in terms of *perfection*. In other words, they are like mathematical asymptotes that represent unattainable but desirable goals – always being sought but, due to human fallibility, never totally attained.

> **DESIRED OUTPUTS**
>
> Performance, results, attained objectives, achievements, and demonstration that assure the **CONTINUATION** of the system's existence.

Figure 4 – *Desired Outputs*

Collectively, Desired Outputs are the rationale for the *continued existence* of the system. Should they fail to be realized to some degree, the system would lose its justification to exist.

There is always a *feedback loop* from Desired Outputs to Known Inputs to incorporate "lessons learned" that, over time, increase the efficiency of the system.

To summarize:

> *A System is a TRANSFORMER of KNOWN INPUTS into DESIRED OUTPUTS — accomplishing that feat by utilizing RESOURCES (personnel, dollars, equipment, facilities, technology, time, and reputation).*

Death as a System

Hopefully, we have agreed so far, that virtually anything can be viewed as a "system." But maybe you're wondering, "Can something as prevalent but frightening as *death* really be treated as a *system*?"

Let's attempt it. There is a greater reason than simple novelty to consider human death as a system, however. No other subject dominates humanity – its existence, values, expectations, and well-being as powerfully as death does.

So human death needs to be examined and evaluated in a *holistic, integrative* manner — instead of the current piecemeal and random fashion which unduly confuses and frightens the public. As Ernest Becker says in *The Denial of Death*, "The fear of death is indeed a universal in the human condition."[2]

Human death is a very complex subject partly because it is an *ongoing process* rather than — as commonly thought – only an *event*. But even so, it qualifies as an excellent system candidate.

To start visualizing it as a system, human death's *known inputs* and *desired outputs* must be defined. Using the criteria of Figures 3 and 4, note the conditions, situations, events, or objectives that are proposed in Figure 5 for human life's Known Inputs and Desired Outputs. Ignoring the normal limit of six or seven for both inputs and outputs is allowed here because all ten are required to fully bound and circumscribe human death.

Hopefully, you concur with those inputs and outputs. You might prefer to revise or replace some of them. But let's agree to accept them as sufficiently adequate for our objective of being all-encompassing and systematic about death.

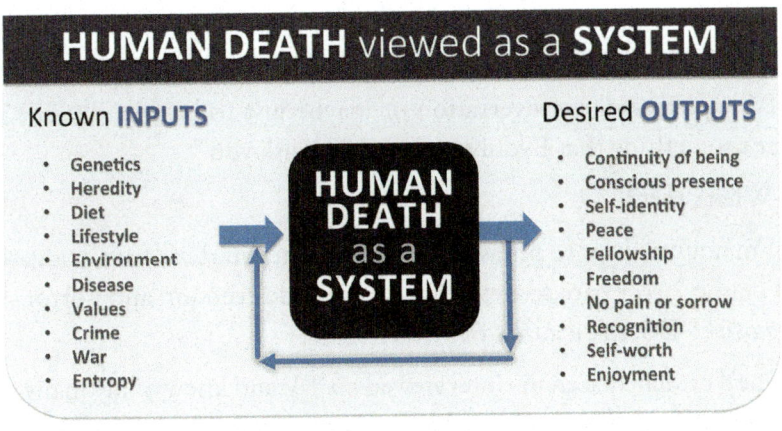

Figure 5–*Death As A System*

Now that human death has been bounded and defined as a system, we are prepared to identify and consider the *risks* in that system . . .

Chapter 2

The Meaning of Risk

"Come friends, it's not too late to seek a newer world." Alfred Lord Tennyson

We were just standing around in the narthex of a Minneapolis church. The rehearsal for our granddaughter Amy's upcoming wedding was getting organized -- everyone lining up in proper order to begin...

A much younger friend of the family who had come to the rehearsal from Wisconsin and whom I knew only distantly was standing next in line to me.

Perhaps to make conversation or maybe just to be friendly, she said, "There's something that I've always wanted to ask you."

"What's that?"

"I'm not sure how to phrase it -- but how can you *stand* to investigate aircraft crashes? They are scenes of such terrible destruction and horror. Isn't it *repulsive*? Doesn't it make you *sick*?"

She'd evidently seen me interviewed on TV and knew of my many years of searching, examining, and analyzing aircraft wreckage to determine why a disaster had happened... Yet no one had ever posed those questions to me.

Could it be that she thought I was weird, ghoulish, or a sicko? Her inquiry was a bit off-the-wall... but I did begin to wonder... what *does* motivate me? As I reflected on how best to answer her, a quick response came to mind.

"First let me assure you, I never get accustomed to investigating catastrophes -- and I would never seek out such an opportunity. I'm not blasé about them," I began.

"It's always disturbing to probe the details of a scene where people have died suddenly, unexpectedly, and generally under devastating physical forces. I frequently ponder what their last moments must have been like, how much time -- if any -- they had to realize the finality of their life. Often, I cannot eat for hours after witnessing such evidence."

She seemed somewhat relieved to hear that confession. However, I knew that it didn't fully answer her questions. So I continued to explain.

"But, in contrast, I always carry within me -- almost like a stabilizing gyroscope -- a strong sense that I am doing this work to make certain, if possible, that it never happens again. I have become a passionate advocate for the deceased, hopeful that I am taking up the cause they undoubtedly would have pursued had they survived."

Her expression brightened with the revelation of my positive reason to be involved in such tragedies. So I continued . . .

"And that hope is built on an extensive history, involvement, and even success in helping to revise the circumstances which led to previous similar tragedies. In a sense, I've been blessed to be able to learn from analyzing many horrendous accidents what must be done to reduce their probability in the future."

Not All Altruism

My response to her honest questions, however, may strike you as too self-assured — even pompous perhaps. And you might be right. Without any doubt, my personality does not always reflect the devotion and concern for others that I would like.

My wife, Phyllis, is a nurse with so many admirable traits of mercy which I lack. And I often wish that I shared those qualities.

Though a controlling, power-driven, Type A person who would not hesitate to "take charge" in many situations simply because I feel as qualified as anyone else around, I'm also aware of my inadequacies. But even my presumptive drive never lets me take my eye off the goal: *doing something to stop a tragedy from happening again.* That's always my focus.

In retrospect, my brief impulsive answer to my friend's surprising questions may or may not have been sufficiently adequate for that moment.

If I were to add to it, however, I would acknowledge my desire to repay a substantial debt of gratitude I owe so many who have taught me throughout my life many lessons worthy of perpetuity.

History as Teacher

On one side, every catastrophe involves *known precedents* — that is, similar or identical factors and aspects that have occurred in the past but failed to be sufficiently corrected and prevented. Thus no accident is totally new.

On the other side, since there has already been so much work devoted to accident prevention, it becomes a real challenge to conceive unique actions to implement that either (a) have never been proposed or (b) deserve reconsideration even though they have been rejected in the past.

All major disasters — not just aircraft crashes — are very rare events indeed. *Why* are they rare? It is partly due to *previous* similar tragedies — and the subsequent corrective actions taken to preclude another identical one. So *safety progress* might be defined as "a long chain of cumulative actions taken to overcome and correct deficiencies as they have arisen over time."

The key to preventing a calamity like an aircraft crash lies in broad knowledge of all the factors that enabled it to occur. At the outset, one must know or be able to reconstruct what happened along the way to the tragic event. Those "happenings" must be viewed within a global, all-encompassing framework. That's where the systems approach proves to be essential.

Then each of the enabling factors — whether involving weather, aircraft operation, human behavior, scheduling, communications, aircraft design, airline management, air traffic control, and so on — become candidates for correction, either by being *removed* or so *modified* that they can never again contribute to such a loss.

As a Member of the National Transportation Safety Board (NTSB), it was my responsibility to head investigations of disasters as shown here. Every crash required a team of specialists possessing expertise in those enabling factors to go on-site where they probed for clues to what had failed … to hopefully use that knowledge to propose, discuss, and recommend actions to avoid repeating the disaster.

The Meaning of Risk

Examining – with always present security and news reporters – wreckage from a mid-air collision occurring near San Luis Obispo, CA on 17 August 1984. Shortly after a commuter airliner took off, a small private aircraft collided head-on with it.

The systems approach provided a rational and thorough framework for assembling and analyzing all the information gathered on-site to derive probable *causes* (there are always multiple causes) leading hopefully to countermeasures that will preclude future similar tragedies.

RISK – Foundational Forcing Factor

Recounting the progressive experience of ascertaining *causes* of disasters reveals a pattern which can be seen only when we recognize its universal stimulus – a term that everyone recognizes and has experienced. We call it **risk**.

If *death* is the WHAT, *risk* is the WHY. And there are almost innumerable reasons why people die. Dallas Cowboys chaplain Howard Hendricks had a famous saying:

"Some men die in ashes, some men die in flames. Other men die inch-by-inch while playing silly games."

Is it not ironic that football fans by the thousands scream in the stands, "Kill 'em!" until the same player is gravely injured – whereupon they immediately begin praying that he is not *fatally* injured?

Video games encourage youngsters to create or manipulate death with impunity. Of course, there are innumerable contests of all kinds and forms that we all play with death – teasing and taunting risks. That age-old maxim "Nothing *ventured*, nothing *gained*" implies that *venturing* involves some degree of risk — with risk simply being the price for getting good things in life.

However, the primary reason that commercial aircraft crashes arouse such wide public interest is because they generally involve massive, sudden, unexpected death and destruction — well beyond the daily personal level of individual risk-taking.

Easily – even subconsciously – anyone who flies aboard an airliner correlates every reported crash with their own perceived *risk beyond any possible personal control*. The moment that airliner door slams shut for takeoff, all *personal* risk is subsumed into one *collective* risk of a crash.

Even though the probability of an airline crash is *one in every 19, 000 flight-years*, it is one of the six most feared causes of death.

Such disasters initially appear mysterious – raising immediate public demand for a *raison d'etre*. Television news explodes with sensational, speculative probing of expert opinion — often even my own. Having given over 500 such interviews, I have frequently participated in that feeding frenzy of wondering "what, why, and how."

The prime focus is always on recovering the aircraft's flight data recorder (FDR) and cockpit voice recorder (CVR) that help to answer the mystery of *why*. But recorders cannot account for such factors as pilot competence, aircraft design, visibility, crew awareness or fear, distractions, cockpit lighting, crew bodily movement and gestures, or external visual stimuli.

So it often takes up to a year of investigation, analysis, interviewing, and deduction before probable causes are resolved and published. Thereafter, risk of commercial airline travel is hopefully reduced – but obviously never *eliminated* – by implementing recommended preventive actions.

Defining Risk

Most dictionaries have multiple definitions of risk. All of them imply peril, danger, hazard, or loss. The entire insurance industry is built around those fearful definitions – occupying one side of life as *insurer* with everyone else on the other side as the *insured* – both sides betting on the probability of loss.

My good friend and respected authority on managing risk, Felix Kloman, advocates defining risk as in Figure 1 because it does not imply *loss* per se. Thereby, risk holds the possibility of becoming an avenue for unexpected *success* as well as *loss*.

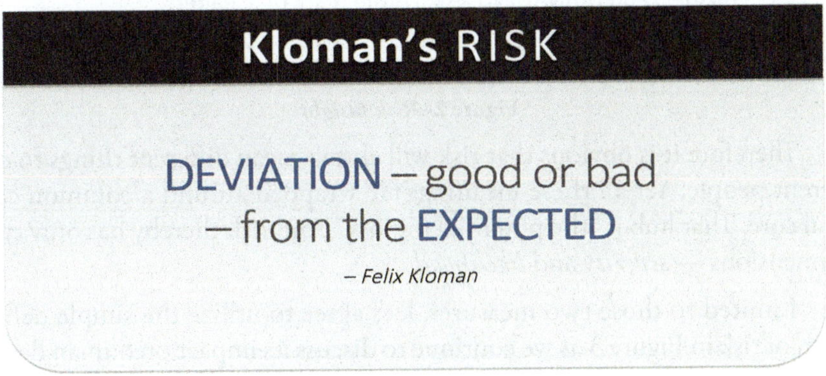

> **Kloman's RISK**
>
> **DEVIATION** – good or bad – from the **EXPECTED**
>
> – Felix Kloman

Figure 1–*Kloman's Risk*

However, *managing* Kloman's version of risk would be a real challenge. What *units* would *you* use to measure "deviation"? How would a *baseline* for "expected" be defined? Even "good" or "bad" are vague and ambiguous terms whose meaning has to vary from person to person.

Bankers Trust, a historic American banking organization, once offered the insight in Figure 2 that seemed to support Kloman's concept of risk – yet without *defining* risk!

> **RISK INSIGHT**
>
> - Some risks are clearly visible. Others hide from sight.
> - The unexpected is the one thing you can always expect
> - Life can never be risk-free; leadership isn't built on sure things.
> - Hide from risk... and you hide from its rewards
> - To prosper, you have to stick your neck out
> - Risk and reward travel side by side; avoid one, and the other will also pass you by.
> - But your choice of risk is critical. Some risk you want to take... some, you don't.
> - NOT taking risks... may be the biggest risk of all.
> - Taking... and MANAGING risk... is the mark of a leader.
>
> — Bankers Trust

Figure 2 – *Risk Insight*

Therefore it is obvious that risk will always mean different things to different people. Yet all these meanings are wrapped around a common central core. That hub is "the potential for loss." And risk thereby has only two dimensions — *severity* and *likelihood*.

Limited to those two measures, let's agree to utilize the simple definition of risk in Figure 3 as we continue to discuss its impact on human death.

> **RISK**
>
> The potential MAGNITUDE of a **LOSS** combined with its LIKELIHOOD

Figure 3 – *General Definition of Risk*

Twins: Risk and Loss

As illustrated in Figure 4, wherever risk exists, its Siamese twin is *loss*. And the loss of our concern is human *death*.

Vital **RISK-LOSS** Linkage

- RISK **precedes** LOSS
- RISK **warns** of LOSS
- RISK is not easily **measured** but LOSS is
- RISK Assessment **estimates** LOSS potential
- RISK Management **minimizes** LOSS
- RISK specialties all aim at **reducing** LOSS
- LOSS **unifies** all sources of RISK

Figure 4–*Vital RISK-LOSS Linkage*

It is interesting that death is often expressed as "something lost." Loss of life, lives lost, and losing life are terms that link death with a *possession* that can be taken away from us voluntarily or involuntarily.

PEOPLE and Risk

Universally, people are entranced with risk — not unlike riverboat gamblers. We flirt, we tease, and we get excited by taking risks.

It has been statistically demonstrated that we openly accept *1000 times more risk* — if we do not have to surrender control to someone else. The long list of examples of this wild disparity include ... smoking, skydiving, driving after consuming alcohol or drugs, bungee-jumping, skiing, gambling, and hunting. Whether personally or vicariously, risk attracts our attention like the proverbial "moth drawn to a flame."

Risk has many faces — threat, hazard, falsehood, liability, environmental pollution, financial bankruptcy, and toxicity. However frightening these faces may be, they all point to a single ultimate event — *human death*.

Because risk is present in so many aspects of life, a wide variety of endeavors, organizations and specialists have been established to manage it.

Each one of these efforts has become a literal domain or fiefdom – with its own language, expertise, advertising, techniques, specialists, organizations, and focus on risk.

Nine such risk specialties are shown in Figure 5.

Figure 5–*Risk Specialties*

The irony of this specialization is that each of these pursuits has evolved into becoming a *competitor* rather than a partner – every specialty convinced that they are *exclusive* experts in managing and controlling risk when they are only one of many efforts attempting to reduce it.

On the other hand, those nine kingdoms illustrate the *magnitude of effort* that has arisen and been organized to address and control risk. Collectively, they constitute a multi-billion dollar industry!

Thrill is almost always linked to *risk*. Yet, it is obvious, when examining the risk of human death, that its ultimate *likelihood* is 100%. No one escapes death. So why is it thrilling?

On the other hand, the *severity* of human death's risk (i.e., the process, timing, and conditions of dying) is where all the focus, variability, and interest reside — because it offers the single option for *delaying* the ultimate. Only by reducing risk's immediate severity— via medical science, protective clothing, law enforcement, and other means — can human death *temporarily* fade from concern.

Hopefully, we can agree that managing the risk of human death must be focused exclusively on reducing its *severity* – never on its inevitable *likelihood*.

Identifying HUMAN Risk

Recall Debi's frightened exclamation, "I'm not ready to die"? Why did she suddenly say that? It was aroused by a situation – actually *an identified risk*. That risk was the suddenly realized possibility that she and her husband – flying together on a charter airliner – could simultaneously die in a crash, leaving their children behind as orphans.

Prior to my conversation that day with her, however, she had never considered or even been aware of that risk. It was *unidentified* – for her. Yet it was there all along!

For every one of us, almost *all* risks we face are and remain unidentified. We live with them unconsciously. So to *identify* risks, we must "Start at the very beginning – a very good place to start," as Julie Andrews sang in *The Sound of Music*. At the outset, we must recognize that all risk involves two aspects that affect human death: *depth of its severity* as well as *breadth of its possibility*.

Humanity's Two Risk Dimensions — Magnitude and Likelihood

It should be obvious that any given risk may impact you on a scale from minor to disastrous. Importantly, any personal risk on that scale must be *identified* before you can *manage* it. Otherwise, it will catch you unaware. And if it is sufficiently severe, it will kill you.

We've already agreed that people die due to a myriad of very different and diverse risks. We can likely further agree that – consciously or subconsciously – we all try to avoid even *thinking* about those risks. Therefore, what is needed is a simple technique to start identifying human risk.

One approach to identifying risk is to treat the anatomy of a specific loss or risk as a flower. Figure 6 illustrates the idea that loss from a risk may take time to develop – similar to a *flower*. Its source may be out of sight, like a hidden root. It is enabled and bolstered by societal factors that are often obscure and subtle – giving no clue of its ultimate appearance as it grows.

Then suddenly a loss bursts forth like a flower – generally as a surprise and without warning.

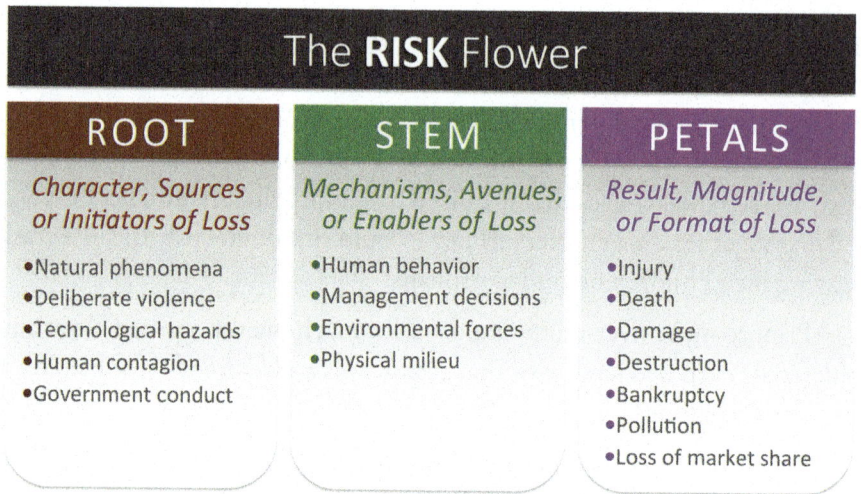

Figure 6 – *The Risk Flower*

The subtlety of risk development frequently disguises its predictive potential prior to loss. So it is important to recognize that a developmental process – after a loss — facilitates creation of effective preventive countermeasures following any loss.

Employing this metaphor is intended to stimulate imagination about the diversity of risk since most of us tend to focus too narrowly – if at all— on its potential for loss.

Risk Significance Concerns

There are a variety of risk issues that influence the frequency, probability, and severity of death potential. *Ignorance* of risk – whether deliberate or inadvertent – may be the most contributory to death. *Aging* obviously is a factor because physical strength, visual and aural acuity, and awareness degenerate with age. *Responsiveness* to stimuli also deteriorates in later life – resulting in failure to avoid obvious risks. Even *awareness* of hazardous situations and conditions – after their occurrence and wide disclosure — frequently decreases as time passes following a disaster or catastrophe.

The statue on the right side of the main entrance to the National Archives of the United States in Washington DC bears the inscription "Eternal vigilance is the price of liberty." Such ongoing awareness is actually **a *mental state concerning risk*** that unfortunately atrophies over time. Yet if that price is not paid, the potential for death is obviously increased.

LIVING With Human Risk

By now, it should be evident that – for humans — risk and death are tightly linked. So the challenge is for everyone to reduce or control risk in order to delay death's arrival. But in practical terms, how can someone actually *diminish* risk?

The idea that an *individual* – living in a complex world as only one of seven billion other human beings – could control the amount of risk that swirls around them may seem almost incomprehensible. On the other hand, to ignore all potential risk would seem to invite compression or reduction of life – accelerating the arrival of its terminus, *death*.

Foreseeable — But *Manageable*?

Wisdom suggests that being aware and searching for risk is an important first step. We acknowledge that it is rarely done in an all-encompassing manner. Most often, it is undertaken after being triggered by experiencing or observing a big loss — personally or vicariously. That effort – regardless of its rigor — is an acknowledgement that risk is *foreseeable* and can be avoided.

However, once risk is *identified*, it must then be *addressed or managed* – if its impact is to be avoided or even minimized. Not many of us are sufficiently disciplined to initiate and maintain action that reduces and controls recognized risk.

That realization stimulated me to write a book on *managing risk* — a best-seller that was even translated into Japanese.[1]

The *Intrigue* of Risk-Taking

None of us are really rational about daily living – constantly evaluating situations, values, tasks, and opportunities for inherent risks that we face. Yes, we know that it would be ideal if we were that sensitive and aware. But somehow, other priorities seem to crowd in and dominate. *"Out of sight, out of mind"* and *"Ignorance is bliss"* both seem to describe our attitude about personal responsibility for risk.

Beyond that *unawareness* of risk, another rather unbelievable human characteristic seems almost universal. The very *idea* of risk often creates immediate tantalizing, exhilarating thought in many minds concerning their personal involvement – but at a distance. They love to watch its *teasing* of death and destruction. The higher the stakes, the greater the thrill!

Why are the Navy *Blue Angels* or Air Force *Thunderbirds* scheduled to scream "on the deck" across crowded stadiums just before kickoff of important football games? What is their connection to the game itself? Could it be that their risky tight formation at such low altitude is intended to compliment the risk about to take place between 22 men on the field? Whatever the reason, the result is to momentarily satisfy human exciting entrancement with risk.

Imagine a circus without demonstrated defiance of risk — by tight-rope walkers or human bodies being fired from cannons. The audience is thrilled vicariously – urging even more dangerous acts! Why is *that*? What is it that is so *appealing*? Could risk-taking actually be an indigenous – even incurable — part of human nature?

EXPOSURE to Risk

Risk is everywhere – seen or unseen. All seven billion humans on earth move about in an environment of risk that exists regardless of personal intent, awareness, involvement, or consent. Humanity is assaulted by risk of earthquake, tornado, tsunami, volcano and other natural phenomena that occur randomly – both in timing and location. Man-made violence – war, terrorism, murder, arson, rape, and robbery — flourishes everywhere on earth without ceasing.

So humanity is both source and recipient of undiminished risk. That universality of exposure and involvement requires that we examine two aspects of exposure in greater detail – thereby enabling some degree of managed risk.

VULNERABILITY — Deliberate v. Inadvertent

Not all risks apply to all people. But even if exposure to risk were to be universal, some people are more likely to suffer harm than others. Why? Because there are folks who deliberately and knowingly subject themselves to risk.

Construction workers building skyscrapers hundreds of feet in the air, skydivers, and medical personnel treating sufferers with the deadly Ebola virus are readily recognized as vulnerable to high risk.

But there are also a large number of other people whose so-called *ordinary* jobs involve deliberately accepted risk. They are aware of the inherent potential of death. Figure 7 lists the Bureau of Labor Statistics' top ten occupations involving risk. And the cost of their insurance policies undoubtedly reflects that high risk.

Ten Most DANGEROUS Occupations

- Commercial Fishing
- Law Enforcement
- Military
- Logging
- Fire-Fighting
- Pilot
- Coast Guard
- Mining
- Trucker
- Roofer

Figure 7 – Ten Most DANGEROUS Occupations

In contrast to those who knowingly accept risking their lives, there is a much larger group of people who are constantly subject to unrecognized risks. Whether due to ignorance, neglect, suppression, or attitude, these folks are unknowingly vulnerable to a wide range of risks.

It is difficult to classify those who deliberately build homes on earthquake fault lines or in hurricane danger zones where storm surge is the leading cause of deaths in the United States from tropical cyclones and hurricanes. Some would attribute it to "herd behavior" —describing how individuals in a group can act collectively without centralized direction. Others might believe that it is due to *ignorance*, deliberate or otherwise. However, "ignorance of the law (or risk) is no excuse."

ASSAILABILITY — Personal v. Societal

Whereas *vulnerability* addresses the onslaught and impact of risk from the *human* perspective – the likelihood or openness of people to risk as an oncoming force, it could be helpful to take another perspective of risk and humanity.

If we were to imagine risk as a *provocateur or agitator* of human loss and assaulting humanity, it might help to provide a contrasting perspective. In this view, risk is on the offensive against humanity – seeking the most effective avenue or means to achieve its objective of loss. That assault weighs attacking humanity at two levels. The first is as *individuals,* the second is society as a *whole*.

We have already postulated that all of us – as individuals — are guilty of ignoring most risks. And it is more than simple ignorance. We prefer to see the bright side of life, to shut out the unpleasant, to pursue what feels good. Even foregoing foods that add weight or regularly exercising to maintain health are thought — by most of us — to be unnecessary. We know better, of course.

So if risk had personality and was planning a strategy for attacking humanity, individuals would appear to be good targets and quite penetrable.

But what about society at large? Is it a better or lesser target for risk to attack?

Beyond the fact that society is a *collective* of individuals, is it possible that it might offer a different probability of success for frightening humanity into recognition of loss – the result of risk? In other words, is risk better understood by individuals or by society?

As we ponder how best to apprise or educate human beings about risk, it seems initially that there is little difference between individuals and society-at-large concerning risk since people are people whether solo or massed. But there are some mitigating factors to consider.

First, we have already mentioned the "herd behavior" instinct where people sometimes do things *collectively* that they would never do *singularly*. War is an example. Nationalism is another. Religion likewise binds people together into group-think. So people do view risk differently depending on the amount of individuality, volition, or choice is available.

Second, education tends to differentiate society based on perceived risk. The more that risk is identified, documented, and distributed, the more those better educated and thereby having access to it will tend to acknowledge and address it.

To summarize the impact of available risk knowledge on humanity, it seems more likely that *individuals* – rather than society at large – are likely to not only be targeted but also respond to it. Technology – with its many venues – has greatly increased the availability of risk knowledge, but its true impact and change in behavior is unknown.

DEATH as *Ultimate* Risk

By now, it should be evident why death is the ultimate risk. No other risk exceeds its significance. No other risk can follow it sequentially. When "loss of life" is mentioned, it is always a terminal event – the end of the line. Death certificates, stone grave markers, and obituaries represent finality.

Many *Faces* But Singular *Destination*

Risk is something odd. And it's deceitful. Its destination is deceptively hidden and disguised. The story of Eve being tempted by the serpent in the Garden of Eden reveals its nature – questioning whether she will *really* die if she eats forbidden fruit while simultaneously offering her an irresistible alternative.

So risk always involves *tradeoffs* – this for that. And whatever *this* is, death is *that*!

Risk also *sells*. It is the sustenance of so much political power – the risks of global warming, nuclear power generation, GMO (genetically modified organism) foods, and militant Islam provide politicians with stimulation for virtually unlimited financial contributions. Risk is loaded on both sides of those frightening issues. Yet the end of them all is human death.

Baseline for Measurement

For many years, I have taught a variety of courses on the subject of risk in universities as well as in seminars. Frequently, I open the first session with a startling and resounding declaration intended to set the stage for lively discussion: "Everyone in this room is going to *die!*"

While that proclamation is often used by terrorists or crazy gunmen, I always explain that it won't happen *immediately* — and that I don't intend to kill them. Instead, the group is invited to list and describe the specific conditions, settings, environments, and surroundings which they would desire for their own death.

The list is never short. It will always include such things as freedom from pain, following a consequential life well lived, surrounded by loved ones, awareness of departure certainty, and having an updated written will. Often it will also include – for those of religious conviction — certainty of destination. Others may also list the desire to have life be the terminus or end of their existence.

What would *your* list look like?

Why *Assess* Risk?

One of the politically popular approaches to dramatic or widely-publicized risks is to announce that a *risk assessment* will be conducted or is underway. Presto! It is almost as though the public should now roll over and go back to sleep – since *something* is being done about such a frightening risk.

Yet seldom is anything ever *done* about that risk! Risk assessment has become a common antidote or delaying tactic for risks that alarm the public – like chemicals in the food chain, mammography for breast cancer, and texting while driving. It is almost a guarantee for burying or silencing a risk of public concern while appearing to be concerned.

Why is risk assessment so useful for political purposes? Because "assess" means to "determine the importance" of something. There is no *finality* in that determination – no action to be taken, no mandatory resolution. It is an ideal means of moving a risk out of public view while doing absolutely nothing about it.

Major risk assessments do nothing more than *evaluate* an identified risk. They stop short of *action* because, as shown in Figure 8, risk assessments accomplish only the first 4 of 6 required steps for *managing* risk. Often it is no more than an expensive ad hoc effort whose ultimate output is a document containing information leading nowhere.

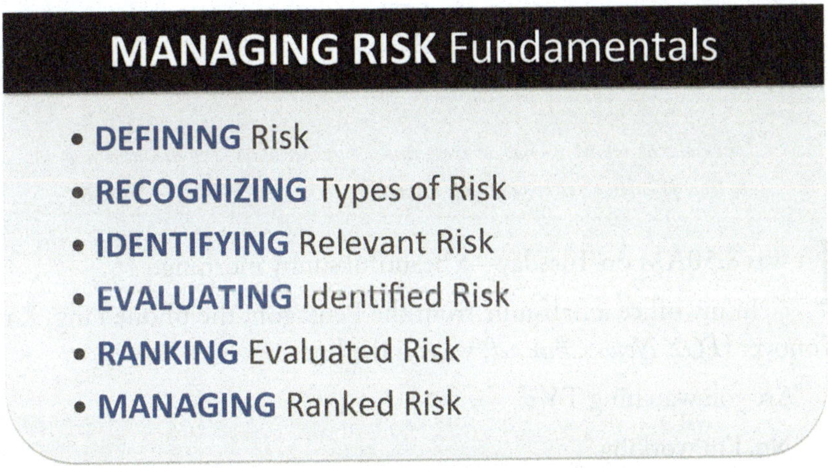

MANAGING RISK Fundamentals

- **DEFINING** Risk
- **RECOGNIZING** Types of Risk
- **IDENTIFYING** Relevant Risk
- **EVALUATING** Identified Risk
- **RANKING** Evaluated Risk
- **MANAGING** Ranked Risk

Figure 8–*MANAGING RISK Fundamentals*

In conclusion, we need to summarize three major points about risk and humanity:

- **Risk is the potential for human loss**
- **Risk is universal – everyone experiences it**
- **The ultimate risk is loss of life or DEATH**

Chapter 3

Human Obsession With Death

"The hour of departure has arrived, and we go our ways–I to die, and you to live. Which is better God only knows." — *Socrates*

It was 8:50AM on Tuesday—a beautiful sunny morning.

In my office a half-mile from the Pentagon, the phone rang. Katie Conover (*FOX News Channel*) was on the line...

"Are you watching TV?"

"No, I'm working."

I was just finishing a report on Air Traffic Control risk for our client NavCanada, second only to the FAA as the largest air traffic control system in the world.

"Well, I'm calling because a small commuter aircraft – possibly a two-engine plane – collided 5 minutes ago with the World Trade Center in New York. Do you think it might be an accident or could it have been intentional?"

My immediate response was to ask her for details about what type of aircraft it was, what direction was it flying, how high it hit in the tower (questions for which she had no answers). I also began reciting factors that could possibly resolve whether it was accidental – high air traffic density around Manhattan due to nearby Newark, LaGuardia, and JFK airports, air traffic control radio contact with the aircraft, unexpected aircraft failure, pilot distraction...

As I continued, Katie interrupted me.

"Would you be willing to go on the Network in New York right now by phone and share your observations?"

"Sure, go ahead."

Shortly, Jon Scott who was the FNC host at that moment – was introducing me, mentioning my background as a former Member of the National Transportation Safety Board. It was now around 8:57. Jon had often interviewed me. And he's a pilot.

So we began discussing air traffic density in the New York area. He was obviously seeking some rationale as he watched on a monitor the horrific fire that had erupted in Tower One.

To support the possibility of an accident, I had just started to describe the collision in July 1945 of an Army Air Force B-25 medium bomber into the 79th floor of the Empire State Building in a thick fog, as it was attempting to land at Newark.

The transcript of our discussion follows . . .

SCOTT: "That, as we've said, was a fog . . . that was an accident, wasn't it fog . . ."

GROSE: "Yes."

SCOTT: ". . . that obscured the pilot's vision?"

GROSE: "Yes, it was definitely an accident."

SCOTT: "Can you think of any reason for a pilot to slam into a building of this height on a day like today, if it wasn't intentional?"

GROSE: "Well, there could have been some inattention. I wouldn't rule that out. In other words, he might have had difficulty . . . might have had engine failure. He might have had his head down in the cockpit instead of looking where he was going. And it depends also on the angle of the sun, for example. Was he flying east when he hit? I don't know because I can't see the pictures."

SCOTT: "Even for light planes, there is a traffic way up the Hudson River not very far from these buildings, isn't there?"

GROSE: "Oh yes. The buildings are close to the River. And furthermore, you've got three major airports in the area – LaGuardia, Newark and JFK. And it's the busiest air corridor in the world, from that point of view. So he could have been confused."

SCOTT: "And there are some height limitations. In other words, the planes that fly by there – I'm talking about personal aircraft, light aircraft, and even some, I guess, commercial aircraft. I know that I've returned to the city from time to time and come in at about . . . about the height level with the top of the building there."

GROSE: "Well, certainly. And you have a lot of helicopter operations in that area down at that low altitude."

SCOTT: "So is it, is it in your view . . . this could have been an accident. We're not necessarily talking about a deliberate act here."

GROSE: "I wouldn't think so immediately at all. You've got to find out why he crashed. Obviously if it's a deliberate terrorist act, then there'll be a message probably from somebody taking credit for it. But right now, it sounds as though the early morning sun . . . we've had drivers in cars get blinded, you know, by the sun. You could have aircraft trouble. So he could be distracted. There are a number of reasons why the aircraft might hit the building."

SCOTT: "Eyewitnesses are saying it was a small commuter plane. Again these early reports are just coming in now from the wire services . . . a small commuter plane apparently hitting the side of the World Trade Center. Happened just a few minutes ago, but you can see the smoke, the smoke tower is growing. There is quite a bit of flame inside the building. The two towers are home, at least during the day, to upwards of 50,000 workers."

GROSE: "Yes, I understand that."

SCOTT: "I'm reminded of a couple of things that happened recently, Dr. Grose. Not long ago, within just the last couple of weeks in fact, there was a pilot who flew . . . (*at this point, Jon gasped – he was watching a monitor as United Airlines Flight 175 curved and smashed into Tower 2 at 9:03*) . . . **there was another one!** We just saw . . . we just saw another one. We just saw another one apparently go . . . another plane just flew into the second Tower! This raises . . . this has to be deliberate, folks."

GROSE: "Well, that would begin to say that. Yes."

SCOTT: "We just saw on live television, as a second plane flew into the second tower of the World Trade Center. Now, given what has been going on around the world, some of the key suspects come to mind – Osama bin Ladin – who knows what. . ."

That abruptly ended the interview!

Katie called right back on the phone and asked if I could come in quickly to the FOX studio in Washington. I told her to order a limousine and let me know when it would be there. She responded immediately — it would be there at 9:30.

When I went down to the lobby at 9:27, the limousine was already there. My Pakistani driver drove out of the lot, passing by the Pentagon. Due to his very fast and daring driving through lots of road construction, we got to the FOX studio in Washington at 9:37.

That was the moment that American Airlines Flight 77 smashed into the Pentagon!

When I walked into FOX, it was pandemonium! Everyone was standing and staring at a row of overhead TV monitors tuned to many different channels. I looked at one monitor and saw that the Pentagon was on fire! I couldn't believe it, as I had been there only a few minutes earlier!

People were running everywhere saying that this was war! Many were discussing with each other, "What does a TV channel run, if war has broken out?"

I was rushed onto a set – wired with a mike and earpiece to begin listening to network audio in NY. There was a TV monitor in my darkened studio, so I could watch what was happening. Obviously, I'd never seen anything quite like it.

The most sickening to me personally was watching the collapse of Tower Two. It was surreal! For our 49th wedding anniversary the previous year, Phyllis and I had gone to New York and had made our first visit to the top of the World Trade Center that had now disappeared!

In the 1970's when I was teaching at The George Washington University, one of my students was the Safety Director for the Port Authority of New York that manages the World Trade Center. He had explained to me that, on an average day, there were about 40,000 people occupying both towers.

He also said that the average emergency egress time out of those towers was *3 hours*. On every 5th floor, he said that there was independent fire suppressant capability – I understood it to mean that they had water and hoses available to knock down a fire, since fire fighters couldn't possibly bring that equipment to great heights.

So watching the collapse of those two towers seemed to portend to me upward of *40,000 fatalities in one instant in one location!*

Conspicuous Ghost Outside the Door

Death is an ever-present possibility – wherever we are. To imply that it lurks in shadows around every corner may be a bit too poetic. However, we all keep it at a distance – until it overpowers us. That was what occurred on September 11, 2001 for America.

Even though both the World Trade Center towers had now collapsed, I continued to sit in the darkened FOX set in Washington, all "miked up," listening to audio reports in my earpiece, staring in unbelief at the monitor, and wondering when they would be coming to me for an interview.

Frightening reports kept coming in — one after another. One said that there was an aircraft approaching Washington from the south, due in 10-20 minutes possibly to hit the Capitol. Sirens began to wail wildly outside the building, but I had no window. The Capitol, only a block or so from us, was ordered to be evacuated. Thousands began pouring out into the nearby streets...

All of the terror reports later proved to be false. After about 45 minutes of waiting, I was joined on the set by Steve Pomerantz, retired Assistant Director of the FBI for counter-terrorism. We sat there together for about an hour – still without being interviewed. He and I mulled over what we were seeing on the monitor. I expressed thankfulness that, so far, there had been no biological, chemical or nuclear assault. However, we both pondered what point of attack might be next... Chicago, Los Angeles, Atlanta...

Meanwhile, I asked for a phone to be brought on the set so I could call my wife. I was concerned about how she was doing and wanted to know what she was seeing at the Pentagon that I knew was now burning only 800 yards away from our condominium penthouse.

Phyllis told an incredible story of hearing the American Airlines 757 approaching at 540 mph – forming a powerful ground wave that shook the entire building before whooshing by terribly low (below our level and about 150 yards north of it). She had run outside on our patio and watched just as it smashed into the Pentagon in a huge ball of fire.

Not immediately associating the aircraft with the massive fireball, she thought that it might have been a bomb dropped by a low-flying aircraft that flew on. But the aircraft itself was the "bomb." Her first reaction was, "World War III!"

As soon as she began sharing her frightening account, I urged her — aware of the historic significance of the crash — to begin shooting both photos and videos. The Pentagon burned — on and off — for 3 days!

The first photo of a series taken by Phyllis Grose from our penthouse overlooking the Pentagon — about 30 minutes after impact of American Airlines Flight 77 into its SW side on 9/11.

Meanwhile, I sat on the FOX set awaiting an interview. As I watched the scene in New York unfold, I recalled a Biblical excerpt — I Thessalonians 5:2-3: *"You know very well that the day of the Lord will come like a thief in the night. While people are saying, 'Peace and safety,' destruction will come on them suddenly, as labor pains on a pregnant woman, and they will not escape."*[1]

The next day, when I read that Minoru Yamasaki who designed the World Trade Center had said "The World Trade Center is a living symbol of man's dedication to **world peace**," and recognizing that the Pentagon was likely the highest symbol of **world security**, that passage seemed even more relevant.

Steve and I finally moved out of the set as Oliver North waited to replace us.

Ollie and I had known each other — professionally as well as personally — for years. He engaged our firm to systematically analyze risks of bulletproof body armor produced by his company, Guardian Technologies International.

My associate Mike Murphy and I discussing risk countermeasures for body armor with Oliver North in his office on 5 September 1996.

Ollie broke the sad news, as we passed each other, that author & TV commentator Barbara Olson had been on American Airlines Flight 77 – and was killed as it smashed into the Pentagon. He asked me to be a guest on his afternoon radio show at 3:30 PM, if I was still at FOX.

Interviews remained chaotic throughout the day – primarily because "breaking news" kept breaking into scheduled interviews. So I ultimately joined Ollie – and during his radio show, he described how Phyllis had witnessed the Pentagon explosion that morning.

Never-Diminishing Magnitude and Wonderment

During all the confusion, FOX had asked me while I was on the set to sign a contract of availability for the next 7 days. So even though I'd not yet been interviewed in the studio, I went into the guest's Green Room to remain available. It is a very small room seating only about six.

Alexander Haig (Reagan Secretary of State) joined us shortly, as we waited. He was very upset that James Baker, GHW Bush Secretary of State, had opposed killing Iraq's Saddam Hussein during/after the Gulf War. While he was still there, reports came in that current President Bush was flying to Barksdale AFB in Louisiana instead of returning to Washington from Florida. Haig was adamant that it was a mistake for Bush not to come back immediately – showing fear of terror rather than needed confidence for the American people. Later, it was revealed that Air Force One was suddenly diverted to Barksdale — while airborne to DC — when several unidentified aircraft were spotted.

When Haig left for an interview, former Speaker of the House Newt Gingrich came in. From the moment the two World Trade towers collapsed, I had continued to believe that there may have easily been 20-30,000 fatalities. Newt and I began to discuss that figure. I proposed that today there may have been more fatalities than on any other day in American history.

Some TV commentators were comparing today's attack with Pearl Harbor, but I recalled that casualties there had only totaled around 2,500. I mentioned to Newt that, during the Civil War, the greatest death toll in a single day occurred at Antietam. So, knowing that he is a historian, I asked him about the death toll at Antietam, and he thought it was about 23,600. Thus it began to look as though that figure might be exceeded!

Lots of time passed as I waited to be interviewed. Periodically when it seemed I was immediately due to go on camera, someone would come in to touch up my "make-up." One of those times, I looked up in shock as Debi — who had just come on duty as the make-up person — introduced herself to me!

She recognized me right away. Our encounter from CNN days over five years earlier was as fresh as if it had been yesterday. She was still doing freelance make-up work at CNN as well as for FOX. We embraced warmly – so excited to see one another again.

Debi vividly recalled her earlier fear about flying. My stark declaration that she "needed to be ready to die" had also lodged itself firmly in her memory.

"But, Dr. Grose, I want you to know that I no longer have that fear!"

"Really? I want to hear how it happened!"

"When we get a few moments, I'll have to share the details."

She left in a hurry to refresh the make-up on another guest. Breaking news events on that tragic day continued to unfold moment by moment – the crash of United Airlines Flight 93 in rural Pennsylvania, evacuation of both the Capitol and White House, delayed return of President Bush to Washington... So we didn't get a chance for Debi to explain to me *how* she had become "ready to die."

After hearing of the UAL Flight 93 crash, I began assembling data (takeoff times, airlines, departure airports, type of aircraft, flight times, and impact times) for the 4 hijacked airliners. It showed a definitely sophisticated and precise strategy. I shared these data with Newt Gingrich, and he was intrigued — as we both wondered how they had managed to pull it all off so successfully.

I took this analysis to the FOX graphics department where I began working to create a camera-ready graphic of the dastardly attack. FOX started using it before my next interview and it was used on camera for many hours afterward – even though I never got to use it for any of my own interviews.

Senators Breaux (LA), Hegel (NE), and Inhofe (OK) also spent different times in the Green Room with us, as they rotated on camera. So many breaking stories poured into FOX that I soon recognized that I was now a "prisoner" in DC — as all bridges were closed and I wouldn't be able to get home. I enquired about my limousine driver and was told that he was standing by for me, regardless of how long I was there.

All FOX Washington interviews were controlled in New York, and the crazy pattern of "breaking news" preempting other "breaking news" meant that there was no way to know when any of us would go on camera. At 5:30PM when I heard that all bridges were opened to outbound traffic, I decided to go home – with 3 scheduled interviews promised later in the evening.

As assured, my Pakistani driver was standing by the limousine at the curb as I left the studio. He was very sober, driving slowly – in contrast to this morning – even though the city was like a ghost town. I wondered what he was thinking, since his English was very limited. He was still my driver when I returned to FOX at 7:30PM where I had two more interviews – the final one at 1:10AM!

September 11, 2001 – the day of America's surprising onslaught of death — ended for me when I crawled into bed at 3:00AM. I had been awake for 21 hours!

Debi's Resolution

The terrorism that erupted on Tuesday, September 11th continued to demand my time because I had signed the FOX contract of availability. There were frequent interviews for me on Wednesday and Thursday. But Debi's schedule for applying make-up didn't coincide with any of my appearances.

On Friday afternoon, I was again at FOX for an interview. The CVR (cockpit voice recorder) and DFDR (digital flight data recorder) had been recovered from American Airlines Flight 77 in the Pentagon. FOX anchor Linda Vester in New York was interested in learning from me what type of data might be available and how it might be used to understand the terrorists' technique in hijacking the aircraft.

What a delight it was to once more find Debi on duty. The frenzy of TV interviews had also slackened somewhat by then. So after she got me "made up," she joined me in the Green Room as I waited to go on camera.

We were both alone – a rarity during that week. It was a good opportunity to learn the details of how she had gotten over her fear of death, so I reminded her that she was going to tell me about it.

"Well, a few months ago, my grandmother died," she started explaining. "And I spent quite a bit of time at her bedside caring for her during her last days. It gave me time to think a lot about what death is all about. Believe it or not, it was during those last days with her that I came to understand the importance of personal faith. Obviously, if death is not to be feared, there has to be a *reason* – because it's certainly natural to fear it."

"So how did you actually overcome that fear – was it sudden or gradual? Was it somehow related to your grandmother's final days?" I was still curious.

"It really revolved around the issue of whether death is the end of everything or simply the beginning of a new phase of living," she said. "And that's where faith came in."

I continued to probe – especially about what faith means to her. After all, we all exercise faith about so many things – weather forecasts, airlines schedules, email addresses, retirement plans, and health insurance.

"When you say *faith*, Debi, what do you mean?"

"Obviously, I'm talking about faith in *God*. But beyond believing that God *exists*, I realized that it was the resurrection of Jesus Christ that we celebrate every Easter that I had to affirm in my mind – to believe really happened. After all, if He was raised from the dead by God's power, then there is life *after* death. That realization did it for me. I was no longer afraid to die but, in a sense, I had already passed from death into life!"

She assured me that she was not eager to die, of course. In fact, she and her husband were going to take another trip together – without the kids – in the next couple of weeks. Yet she was now "ready to die" whenever her time to depart came around.

Impact of Its *Shadow* Seen Everywhere

Death has no identifiable physical substance. Yet it has a shadow. How can this be? Because reality is far larger than physics and chemistry. It embraces imagination, fear, worry, suspicion, and distrust.

Since 9/11, that unbelievable but true shadow of death that suddenly descended on America has remained. While many in Muslim countries applauded it as deserved, millions in most other countries were aghast at its disruption of normality and killing of thousands. That shadow forced many societal changes that will remain as permanent necessities.

Commercial air travel was greatly impacted – with emergence of a whole new cottage industry of airport security making all travelers suspects and stripping them of their privacy. Such blessings of technology as private wireless communication were usurped surreptitiously by government in the name of protection against terrorism.

Progressive forfeiture of previous freedoms marked the political impact of instituting new forms of precluding or minimizing death potential. New laws, regulations, and standards – all ostensibly required to protect the public against agents, conditions, situations, and threats – suddenly rose up and flourished to trade away freedom for security. Why, and on what justification? Because the shadow of death had encroached into new, previously undetected territory.

Bottom Line: *TERROR*

War has always been a major source of death – since its objective is to subjugate and ultimately defeat opponents by killing them. Progressively, the techniques and weaponry of warfare have increased in lethality. At the same time, combat has become less personal as "weapons of mass destruction" or WMD have proliferated. That makes death more threatening, since it may well be an unannounced surprise, with no opportunity to prepare for it.

I vividly recall 9 August 1945 when my hometown *Spokane Daily Chronicle* that I delivered in the neighborhood had *red* headlines for the very first time! They reported that the warhead plutonium for the second atomic bomb – dropped that day over Nagasaki, Japan – had been manufactured in nearby Hanford, Washington.

As a teenager, little did I realize the significance of atomic bombs as the forerunning ultimate weapon of diplomatic blackmail — threat of nuclear warfare — that continues to this day. What makes nuclear warfare so powerful and threatening? The ability to bring death on an unannounced, massive, and impersonal scale.

Ever since the discovery that people can be influenced by intimidation, terrorism has existed. But what is relatively new in history is its world-wide *dimension*. The primary significance of the 9/11 attacks was to raise terrorism to *international* status.

In his book, *CRUSADERS, CRIMINALS, CRAZIES: Terror and Terrorism in Our Time*, Frederick J. Hacker MD clarifies what has recently emerged as a source of death:

> The resurgence of terror and terrorism is not just a regression to primitive behavior that is deeply implanted in human nature, nor is it merely a return to an uncivilized state erroneously believed to have long since been overcome by humanitarian behavior.[1]

Instead, *death* provides the ultimate leverage in international blackmail. It is always there on the sidelines, awaiting a call. And recall that almost every modern war has "stumbled into existence" by misjudgment, error, or overplaying a guessing game.

What's on the *Other* Side of the Bar?

Famed British poet Alfred Lord Tennyson's final and most quoted poem, "*Crossing the Bar*" was one that my Mother often quoted. He wrote it after a serious illness while at sea and said the words "came in a moment." The metaphor of "crossing of bar" represents traveling serenely and securely from life through death. The Pilot is a metaphor for God Whom the speaker hopes to meet face to face.

Tennyson explained, "The Pilot has been on board all the while, but in the dark I have not seen Him... [He is] that Divine and Unseen Who is always guiding us."

In the world of art, a similar theme is reminiscent of ***The Voyage of Life,*** a series of 1840 paintings by Thomas Cole displayed in the National Gallery of Art in Washington, DC. Cole's renowned four-part series traces the journey of an archetypal hero along the "River of Life." Four stages of human life: childhood, youth, manhood, and old age are portrayed.

Cole describes his painting of old age. "Portentous clouds are brooding over a vast and midnight Ocean. A few barren rocks are seen through the gloom – the last shores of the world. The boat, Man's temporal body, has exhausted its inward force, and floats lifeless on a dead ocean. The hourglass that has measured his time is gone; the Hours, too. The traveler's worldly accumulations have been scattered and no longer have meaning for him. For the first time, his Guardian Spirit appears, standing before him and pointing to the Glorious Light that has suddenly and inexplicably opened through a whirlwind rising nearby from the sea. Angels descend to welcome Man to the Haven of Immortal Life."

That final painting, *Old Age,* is an image of death. The man has grown old; he has survived the trials of life. The waters have calmed; the river flows into the waters of eternity. The figurehead and hourglass are missing from the battered boat; the withered old voyager has reached the end of earthly time. In the distance, angels are descending from heaven, while the guardian angel hovers close, gesturing toward the others. The man is once again joyous with the knowledge that faith has sustained him through life. The landscape is practically gone, just a few rough rocks represent the edge of the earthly world, and dark water stretches onward.

Cole's summary of that scene: "The chains of corporeal existence are falling away; and already the mind has glimpses of Immortal Life."

Irrational, Universal Toll

Thinkers like Freud, Rheingold, James, Zilboorg, Kierkegaard, Jung, and Fromm all share in acknowledging the "terror of death." Certainly no one escapes it. And it is frustrating that there is no way to interview someone who has experienced it – to learn exactly what transpires when it arrives.

The mystery of what — if anything — occurs at its completion must always remain unsolved. As rational as human *life* may appear to be, human *death* has always appeared to be irrational.

Earliest history records that sages have pondered – and declared – what death means to them. Figure 1 illustrates the diversity of their conclusions – all of them with no more factual data or information than you and I have.

Ancient Wisdom on DEATH

"There's nothing certain in a man's life except this: that he must lose it." — *Aeschylus, 525-456 BC, Greek tragedian*

"No one can confidently say that he will be living tomorrow." — *Euripides, 480-406 BC, Greek tragedian*

"Death does not concern us, because as long as we exist, death is not here. And once it does come, we no longer exist." — *Epicurus, 341-271 BC, Greek philosopher*

"No one knows whether death may not be the greatest of all blessings for a man, yet men fear it as if they knew that it is the greatest of all evils." — *Socrates, 470-399 BC, Greek philosopher*

"To fear death, my friends, is only to think ourselves wise without really being wise, for it is to think that we know what we do not know. For no one knows whether death may not be the greatest good that can happen to man." — *Plato, 428-348 BC, Greek philosopher*

Figure 1–*Ancient Wisdom on DEATH*

On the other hand, note that Plato proposes the possibility that death might be *the greatest good* that can happen to us! Could it be that a person's *fear* of death is unwarranted? Quite obviously, there is need for having a basis for reaching any conclusion – good or bad – about death. The reason it seems irrational is that it cannot be subjected to scientific scrutiny – tested objectively for universal repeatability.

Thirst for *Evidence* From Other Side

So the seeming *mystery* of death – and our inability to examine it in a laboratory to determine its consistent properties – creates a deep desire or thirst for irrefutable evidence and proof of its character.

Two recent movies that capitalize on this thirst have had conflicting reviews as to their ability to satisfy it. *Life After Death*, released in 2010, portrays a 16 year-old girl who has lost her twin sister struggling to get along with her stepmother and becomes obsessed with a video game that enables her to communicate with her twin. *After Life* (2013) portrays a cynical young woman who learns to deceive people into believing she can communicate with their dead loved ones.

Both movies suggest that there is life after death and that the dead are recognizable as the persons they were prior to death. But is this any more than wishful thinking?

Seldom *Scheduled* Crossing

One of the disturbing factors about death is its *timing*. As Dr. Ben Aaron mentions in his Foreword, "We all know that we must die, but which of us would be willing to name the day and hour in advance, even if we could."

If we now apply the systems approach discussed in Chapter 2 to your personal life as depicted in Figure 2, you can likely find that you have already participated in at least half – if not more — of those functional blocks. Where are you in that diagram?

And it is obvious that, whether or not you end up being involved in all the blocks, you are destined to ultimately arrive at Block 20 — without any question. It is the universal end of the line.

Human Obsession With Death

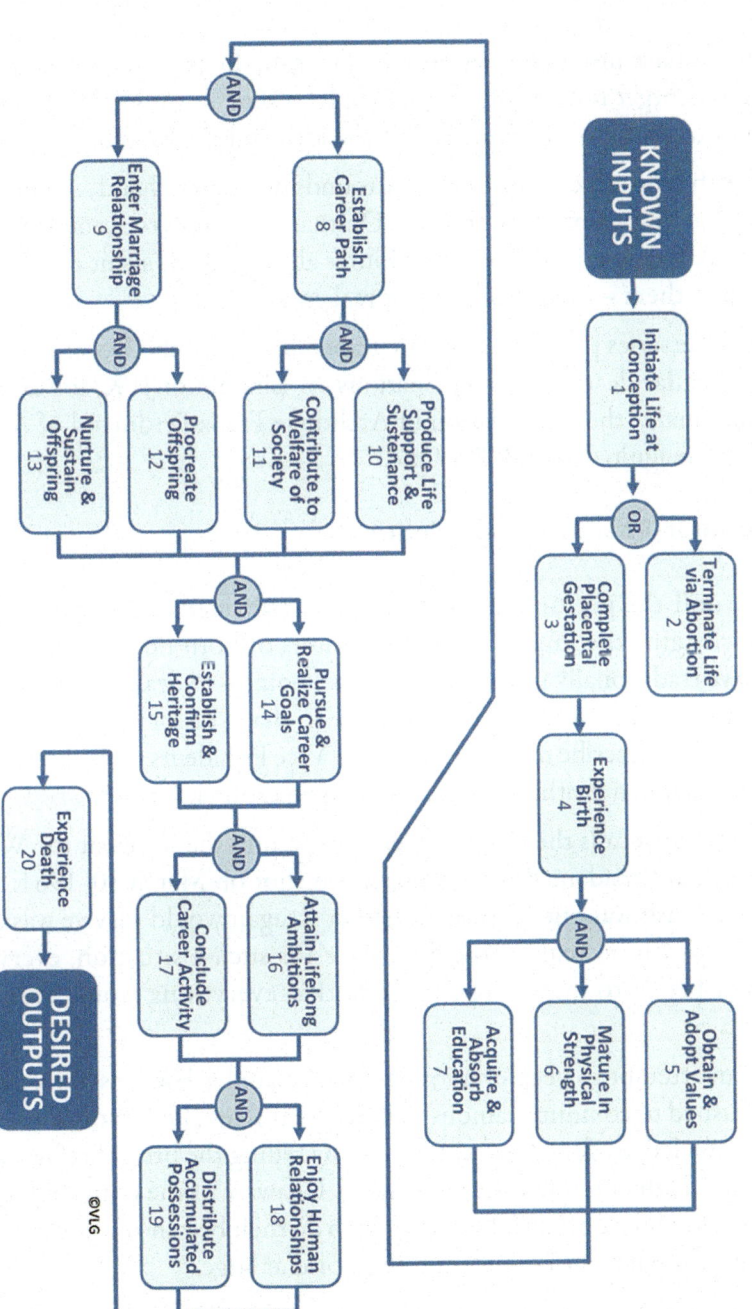

Figure 2 – *LIFE-TO-DEATH as a SYSTEM*

Symbolism Provides a Crutch

We all have invented ways for remembering the past – assuring that we will never forget important events, people, places, or causes. Undoubtedly, you have a means for accommodating the idea of death as well.

Death is no secret, of course. It surrounds us on every hand... reminders abound on how to express its impact. Obituaries attempt to account for lives well lived. Cemeteries – what my Mother always called "silent cities" as we drove past them – occupy prominent real estate.

Anniversaries provide automatic remembrance in the US of particular or spectacular deaths like assassinations — Lincoln or JFK, for example. Internationally, the assassination of Archduke Franz Ferdinand of Austria and his wife ignited World War One.

Footprint in Literature, Arts, Politics

Beyond the personal level, society has provided additional ways of acknowledging the impact of death. Writers of both non-fiction and fiction have traditionally used death as a focal point – to draw the reader into recognition of its overpowering influence to shape history. So much detail is recorded to describe the accession of US Vice Presidents to the Presidency following sudden death – whether by assassination or ill health.

Everyone recalls the chaos that ensued behind the scenes at the White House when President Ronald Reagan was shot on March 30, 1981. Vice President Bush was out of town. Whether Reagan would survive was questionable. With no real protocol in place for such a situation, everyone involved began to improvise and hope that everything would turn out all right.

In an attempt to keep everyone calm, Alexander Haig, as Secretary of State, rushed to commit a famous blunder — and revealed an obvious lapse in his knowledge of the Constitution — by telling the press that he was in charge while the President was in surgery. Unaware of the criticality of the President's condition, key officials began to do their best in keeping not only reporters calm but the country and the world at large.

Antithetical Contrast to *Life*

Any description of death provides a dramatic baseline of demarcation, entirely separating life from death. There is no way to be "just a little bit dead." You either *are* or are *not* dead.

One interesting phenomenon that tried to blur or even deny the impact of death occurred during the nineteenth and early twentieth centuries in American and European culture. Known as post-mortem photography, it was the practice of taking photographs of the recently deceased. Grieving families would engage photographers to arrange and take photos – some with *living* family members, pets, or props – intended to help in the grieving process.

Post-mortem photographs often represented the only visual remembrance of the deceased and were among a family's most precious possessions.

The practice of elaborate preparation of a body for viewing prior to burial is another interesting phenomenon – applying cosmetics, selecting flattering clothing, and elevating the body in a casket – all intended to restore life-like appearance and hopefully minimize the reality of death. Yet the contrast between a living person and a dead person can neither be denied nor bridged.

Personally, this contrast for me was most dramatic one day during the Cold War. I was in Moscow in January. Freezing in a crowded queue for a long winter hour, I waited to view Lenin's cadaver in the Kremlin. It was a ceremony as we entered the tomb — enforced by armed Soviet soldiers at the tomb entrance to assure silence and respect.

As I slowly passed by his remains eerily lighted by a pink spotlight, I noted his clenched right-hand fist. It caused me to wonder what was intended by embalmers in arranging it. Was it to continue a revolution – even in death?

Only *Momentary* Obeisance

Regardless of notoriety, fame, accomplishment, wealth, or authority, death rapidly strips every person of those attainments. Homage may be sustained for a period, but it soon fades. Certainly memory of the deceased is not totally erased, but their influence – with respect to the human race – is diminished by death.

Shakespeare's "Hamlet," Strauss' "Death and Transfiguration," Hitler's holocaust, and Ceausescu's Bucharest bloodbath all perpetually provoke worldwide wonderment about death.

Even death's classical symbols — black color, mournful music, slow cadence, "missing man" formation, solemnity, quietness, riderless horse, halted traffic — depict an interruptive but brief obeisance to its overpowering character.

Recognizing both the brevity and transient nature of life – which is always headed toward death, is it not wise to weigh some deliberate preparation for it?

Chapter 4

Death: The Wrath of God?

"To doubt everything and to believe everything are two equally convenient solutions; each saves us from thinking."
— Henri Poincare

Lightning flashed right outside the left window on a violent, stormy night – lighting up the darkened cockpit of the Avianca Airlines 767 wide-body aircraft.

We were not hit by that flash of lightning, but it was close. Even though I had experienced and survived a few direct airborne lightning strikes during previous airline flights, this bolt of fire was still momentarily frightening.

Flying "jumpseat" directly behind the Captain at 41,000 feet, I am observing the two 767 pilots preparing to descend into busy JFK airport in New York after flying 6 hours nonstop from Bogota, Columbia on 16 July 1997.

Earlier that afternoon, I had met with Avianca Airlines CEO Gustavo Lenis and his executive staff in Bogota to deliver a solicited presentation on managing risk. Once airborne, Captain Carlos Sinisterra invited my associate Mike Murphy and me to join him in the cockpit for the final hour of the flight.

As I listened to radio traffic on a headset, I was amazed that both pilots constantly spoke to each other in the cockpit in Spanish but responded – as required — in English to air traffic control on radio. From my earlier days as an AC&W (aircraft control & warning) controller in the Air Force, I was particularly sensitive to constant air traffic directives among the obviously large number of other aircraft in the immediate area — all lining up in severe night weather to land at 2-minute intervals in New York, like trucks coming down a hill.

In the back of my mind, it was difficult for me not to recall Avianca Flight 52– a 707 airliner that crashed six years earlier into the backyard of tennis pro John McEnroe on Long Island as it approached JFK airport on a similar stormy night and ran out of fuel. In that crash, air traffic controllers had underestimated the seriousness of the low fuel state because the English language used by the Spanish-speaking crew was non-standard and imprecise. Thereby Flight 52 had begun its descent too late to make it to the runway, and 73 souls perished.

The obvious parallel – same huge airport, stormy night, high traffic, same airline, Spanish-English communication — was sufficient to cause me to become very aware of how much my wife and six children needed me as well as how helplessly captive I was to air traffic density, stormy weather, needed orderly descent amongst so many other aircraft, restricted visibility, and alignment for final approach.

Needless to say, I was delighted and relieved as I observed the skillful smooth touchdown on the JFK runway by the Avianca crew.

Everyone *Acknowledges* God

The lightning bolt that night – unpredictable and startling as always – reminded me of man's vulnerability to nature. It is only one of many natural phenomena that have been attributed throughout history to the *wrath of God* — along with hurricane, earthquake, monsoon, cyclone, typhoon, and tornado.

Atheist, agnostic, and believer alike acknowledge God. Atheism is defined as "disbelief in the existence of **God**." Agnostics believe that **God** is unknown and unknowable, while believers are obviously convinced that **God** exists.

Easy Target for *Blame*

Of course, many natural phenomena produce forces and conditions that often result in death. Because we always seek to know *why* these tragic events occur – coupled with our obvious inability to *control* them – these deadly catastrophes have come to be called "acts of God."

There is even a *legal* definition of an act of God: "Any event that directly and exclusively results from the occurrence of natural causes that could not have been prevented by the exercise of foresight or caution." Assessing *blame* – where someone is censured and held responsible for actions believed to be morally irresponsible – is the inverse of *praise*. So God is blamed in the law for doing something that is frightful, devastating, and deadly!

But have you ever wondered why a supreme being should be held responsible for the *inevitable* or *inexplicable*? Even though "acts of God" are admission that humans are powerless to prevent something from happening, why should those catastrophes necessarily be attributed to *God*? And if God is liable, how does that really *resolve* the responsibility issue? Is God any more than a phantom scapegoat?

Evidently, there must be some unspoken but widely recognized rationale for this long-standing conviction. Perhaps it is deduced from the *magnitude* of natural phenomena when compared to human activity. Or maybe imagining a *personality* as an attributable source soothes humanity's own conscience and provides relief concerning guilt for unexpected death.

Rage Must be His Reason

Undoubtedly, *fault-finding* is an essential and required means for settling or resolving the reason for catastrophes. "Coming to closure" – a recent but frequent term for dealing with the impact of unexpected death – often involves identification and punishment of a culprit and the proven reason for their role.

In the case of God as a culprit, the *magnitude* of natural phenomena that far exceed any of human creation has historically provoked an argument for indictment. Thus His personality appears to be fair game as the source of such horrible happenings. Over time, humanity has been enabled by technology to reduce the impact of many natural forces, but it is far from conquering the fallout of all those forces.

As to *why* God might be culpable, the results of natural disasters have traditionally indicated for many a personality of *rage* – fundamentally because people themselves cause destruction when *they* are enraged. In other words, God is transformed anthropomorphically to possess human characteristics.

Vengeful Reputation

God's image has frequently been portrayed in art and literature as judgmental — aligned with vengeance against sin and unrighteousness. Both Old and New Testaments in the Bible quote God saying, "It is mine to avenge." But for what is God's vengeance reserved? For the very violations that go unpunished in so many human domains because of injustice.

Could death resulting from natural phenomena — that are beyond human ability to preclude — rightfully be considered to be due to the *vengeance* of God? On what basis? Since natural disasters so often result in the death of innocent children and the infirm, assigning blame to a vengeful God raises many questions about death itself. Is death punishment? Is death administered without justice? Is death dispatched randomly or specifically?

What Unfair Advantage!

Imputing causes of death from natural phenomena to be "acts of God," raises many questions about the role of God in society.

Attributes frequently assumed for God — omniscient, omnipresent, sovereign, infinite, holy, and eternal — would appear to give Him a distant and uninvolved status on one hand. But on the other hand, those attributes mean that He is *totally* involved in human life! The intimate involvement of Him with humanity would be incredible, if true. How much would He care about who died and under what conditions? What difference would it make that humanity *blamed* Him for certain deaths? To what extent is He *directly* involved in the lives of 7,000,000,000 people alive today?

Assuming that God takes responsibility for all the acts assigned by humanity to be His, what does this mean? He would be in a remarkable and advantageous position to control the time, conditions, location, and means of death for everyone! That is a remarkable advantage.

Massive *Leverage* for Fury

Millions worldwide die every day — many due to so-called "acts of God." Could God actually be involved in — or responsible for — each of those individual deaths? If so, it could be frightening — given His knowledge of all aspects of every human life. Ponder for a moment the vast difference — between a single person and the God Whom most people acknowledge — when defining only *negative* events like death as "acts of God."

The tragedies categorized as "acts of God" are so designated primarily because God is assumed to have control and schedule of tornados, lightning strikes, earthquakes, hurricanes, and all other natural phenomena that wreak fearful havoc on humanity. Yet imagine the *imbalance* between the mightiest human demonstration of power; i.e., nuclear explosion, and the power of any natural example; i.e., earthquake, tornado, hurricane or cyclone. The incredible *leverage* cannot even be measured.

Authoritative Anger

Another common human description of God is one of possessing ferocious anger to unleash against any opponent or situation that provokes His displeasure. This viewpoint reasons that since He has *authority* over all creation, He arbitrarily – even righteously – demonstrates anger through natural phenomena without competition or question.

Irresolvable Role

The final and most significant aspect of designating natural catastrophes as "acts of God" is that God's role is not subject to any superior authority. There is no means or forum for determining or executing revision of jurisdictional authority. Therefore, it is unlikely that God will be replaced as the source of *inevitable* or *inexplicable* acts of nature that result in massive human death and destruction.

21st Century Presence

Even though the perceived or acknowledged role of God in causing natural catastrophes has remained unchanged for centuries, there is evidence even in the 21st century that it will remain as a designated party. The primary reason for its retention is not only that there is no apparent replacement for an omnibus party to blame but the convenience of having a legal category that cannot readily be challenged. What other party, issue or entity can be charged with causing death from natural phenomena?

Inclusion, Not Omission

Death, believed to result from a wide spectrum of causes – from personal to impersonal factors, will almost always provoke a search for its cause. When that search becomes sufficiently sophisticated that it requires legal resolution of causation, it can prove impossible to assign full responsibility for its cause to a person, organization, or party for redress.

It is that dilemma that has led to the classification of an "act of God." That designation allows the death to remain, in a sense, free from *human* responsibility – even when the deceased could have possibly taken action that would have reduced the probability of death; e.g., sought shelter in a massive storm.

This legal assignment of guilt for death will apparently continue to require a *non-human* party to blame. Whether it will be *God* or another entity is not known.

Elimination Awaits Role Resolution

For God to be eliminated as the legal non-human party for blame assessment calls for a *replacement* party as well as a *reason* to replace the current party — God.

Since there is no observable *aftermath* — only *absence* — following human death, God's role in it (should there be one) can never be resolved. There is neither subjective nor objective proof to be scientifically verified that God has a role in death. However, an all-encompassing and systematic approach to human death seems to demand that the issue of God's existence be retained in the equation.

Death's *Causation* Apart From God

We have examined the thesis that it is the inability of humans to overcome or tame nature — as well as the impossibility of eliminating death — that has led to the concept of an omnipotent authority Who uses random, forceful, and hurtful phenomena to maintain an arbitrary dominance over humanity, even to causing its death. That is a harsh conclusion, given that death has the appearance of terminating *life* that has been enjoyed.

What is needed is an alternative to the idea that God causes death. Is death due to a personality similar to God? If so, who is that person? If not, will inexplicable death continue to be considered an "act of God"?

Chapter 5

Demise of Noble Savage Theory

"If we are not ashamed to think it, we should not be ashamed to say it." — Marcus Tullius Cicero

Mike Schere was the best automobile mechanic in my home town Spokane, Washington. And he was employed by my father who owned a thriving automobile business.

As a young boy in the 1930's, I got to occasionally spend a day at work with Dad at Better Chevrolet Service. And I loved it.

Mike was one reason why I had so much fun. He was a character. He chomped on an unlit cigar much of the time. Above his workbench, an Atwater Kent radio was generally blaring out news or music. It was impressive to watch Mike working so efficiently.

One day stands out in my memory.

As I watched Mike busily working, he was also intently listening to the radio. Adolf Hitler was screaming out one of his high-pitched speeches in German. I couldn't understand what Hitler was saying, of course. But I noticed that Mike was really paying attention to every word.

Mike understood. Fluent in German, he seemed very pleased as he listened.

Curious after the speech ended, I asked Mike what Hitler was saying. He said that it was a hopeful speech for Germans. Mike explained that Germany had been in a very bad economic situation ever since the end of World War I, and Hitler was really giving Germans badly-needed *hope*.

Of course, Hitler's version of *hope* ultimately was judged by the world as *evil*.

Evil: Chicken or Egg?

Maybe you're like me. As you grew up, calling anyone or anything "evil" was very rare. True, there were *bad* things. Even terms like *despicable* or *horrible* came up occasionally, but the word *evil* seldom – if ever – was used as a descriptor.

Even the Nuremberg trials – following all the terror that Hitler created in World War II – never used the term "evil." Instead, the court judged German leaders for two types of crime — crimes *against humanity* or *war* crimes.

Crimes against humanity were "particularly odious offences — attacks on human dignity or grave humiliation or degradation of human beings." These crimes consisted of murder, massacres, extermination, cannibalism, torture, rape, and inhumane acts.

War crimes were "serious violations of laws and customs of war involving individual criminal responsibility." Included in this category were murdering, deporting, and mistreating prisoners of war or civilians as well as wantonly destroying cities, towns, villages, or other objects not warranted by military necessity.

Is it not odd that the word "evil" was *avoided*? Why do you think it *wasn't* mentioned? Could its ultimate connection with *death* cause its virtual absence of use? The evil-death linkage was likely assumed without being mentioned.

Marshall B. Rosenberg – creator of *Nonviolent Communication*, a process that "helps people to exchange the information necessary to resolve conflicts and differences peacefully" – claimed that the root of violence is the very concept of evil. Labeling someone as "evil" invokes the desire to punish or inflict pain – even death.

Rosenberg, in his book *Nonviolent Communication: A Language of Life*,[1] said the use of language in Nazi Germany was the key to how the German people were able to do things to other human beings that they normally would not do. He even linked the concept of evil to our judicial system, which seeks to create justice via punishment — *punitive justice* — punishing acts that are seen as bad or wrong.

But where does evil *originate*? What *causes* it to rise up – seeking to kill or obliterate? Why is evil – so obvious – an *avoided* word?

Rousseau's Reversal

Evil – as a word and an idea – has a history. It's a *lengthy* history, too.

Jean Jacques Rousseau — a Swiss-born philosopher, writer, and composer — proposed in the 18th century a new perspective on the relationship of human nature to society. His political philosophy influenced the French Revolution as well as the overall development of modern political, sociological, and educational thought.

Rousseau shared the Enlightenment view that *society* had twisted or perverted *natural man* whom he represented as a "Noble Savage" who had previously lived harmoniously with nature — free from selfishness, want, possessiveness, and jealousy. Society thereby was an overpowering and universal villain.

Simply stated, Rousseau proposed that all people were born free from any evil and became corrupted only by the society in which they subsequently lived. Even though he never used the term to which his name has been linked, his "Noble Savage" theory reversed and directly opposed a long-standing earlier view that the human race was universally corrupted by sin and needful of God's redemption. The theoretical "Noble Savage" lived harmoniously with nature — free from selfishness, want, possessiveness, and jealousy.

Transferring the onus and accountability for evil (and ultimately death) from the *individual* to *society-at-large* has played a large role in political thought and practice ever since. The stark contrast between the French Revolution of 1788-1799 and the American Revolution of 1776-1789 can be traced to the Noble Savage theory, with the French accepting it while the Americans rejected it.

The founding fathers of American government denied Rousseau's premise of the inherent goodness of man. That's why they divided governmental authority into three equal parts to avoid its centralization in one tyrant. Yet there was still widespread belief in the United States that evil (as a source of massive killing) was best combated by *societal solutions* rather than by *individual transformation*.

Non-Existent Noble Savage

Evil and death, of course, are cousins. Their linkage – despite the Noble Savage theory – is undeniable. Therefore, whether or not the Noble Savage theory is valid, death occurs.

The biggest argument against the Noble Savage, of course, is its premise that it has *ever* existed. The term *noble savage* is a stereotype that expresses the concept of an idealized indigene, outsider, or "other" that has not been "corrupted" by civilization, and therefore symbolizes humanity's supposed innate goodness.

In English, the phrase first appeared in the 17th century in John Dryden's heroic play *The Conquest of Granada* (1672), wherein it was used by the son of a Christian prince — believing himself a Spanish Muslim — in reference to himself. However, the phrase later became identified with the idealized picture of "nature's gentleman", which was an aspect of 18th century sentimentalism.

Onus for Evil

What is needed for the Noble Savage to provide a valid, rational explanation for evil in the world? Simply to be proven to have existed *anywhere*. Is it possible that such a person or culture – pure and untouched by society – has ever lived? Even if *one* such person were to be found, what explanation is available for the evil that society has developed over time to corrupt that pure soul?

Reality is a terrible foe – readily destroying beautiful, wishful theories. Following the 20th century's two World Wars, and more recently the worldwide eruption of ruthless genocide by militant Islam, trust in the inherent goodness of humanity has waned. Closer to home, skyrocketing crime rates and bolder public flaunting of law by politicians suggest that proof of the Noble Savage is yet to be found.

Solution: Societal v. Individual

We still need an answer for the widespread evil in humanity. Its role in death exceeds natural causes by far. Establishing *accountability* for evil, of course, is fundamental to eradicating it. Who can be charged with *originating* evil? How did it start?

Once it is proven which party – society or the individual – is responsible for evil, then it will be possible to focus all effort on eradicating it. Until that proof however, we will continue to be puzzled about *how* and *when* evil infected the human race.

Inherent Human Goodness

It is so natural to believe that people – down deep – are really *good*. True, there are some very difficult folks who are trouble-makers. But in the main, we are surrounded by nice individuals who make day-to-day life pleasant. Most of the time...

But this cordiality is not *universal*. Otherwise, how can you explain war, divorce, murder, graft, robbery, greed, theft, deceit, deception, cheating, malice, hatred, spite, brutality, violence, and prejudice? From whence do those *undesirable* characteristics and actions arise? Outer space?

How are we *born*? Are we *inherently* good? If so, how do we learn to be *otherwise*? Is there an *infection of evil* floating around that we contract? Is there a *cure* for evil, once we realize we are infected?

Settling the Dispute

The battle lines are pretty well defined. Advocates for both views about the origin of evil are well-equipped with arguments. There is little ambiguity about the issue either – because evil can be readily observed as existent and plentiful. Its results abound everywhere – in every culture, within every nation, on every continent.

Whether the Noble Savage is the most appropriate representation of inherent human goodness is immaterial. Many still believe that evil is a *learned habit*, forced onto initially pure and innocent folks by an infected society.

Contrasting this viewpoint is a conclusion that humans are born with a nature predisposed to potential evil. This nature is *self-centered*. It views everything from a position of domination. To obtain and retain that dominance, means and measures are employed that may readily be designated as evil. In other words, evil is *second-nature* for these people. It may be suppressed or controlled by education or discipline, but it is always potentially present – ready to erupt at a moment's notice.

Interestingly, the end result of both these viewpoints is widespread evil in the world! The dispute is only over *how* people become evil.

Whence Evil — If Not *Society*?

The fundamental issue to be resolved is not whether evil exists at birth or is acquired from societal influence. Rather, it is the *source* of evil itself.

Evil is recognized by its *result* in humanity. Animals do not possess it. It is vicious, destructive, and insidious. Even if you favor the position that it somehow begins to manifest itself in society rather than individually from birth, the big question is "Where does evil *originate*?"

Evil is a comparative phenomenon – contrasted with acceptable, productive, joyful life. It interrupts – even destroys – peaceful living at unexpected times. Its reality is not disputed.

21st Century Evidence

The influence of Rousseau and the Enlightenment movement did not last. In the 19th century, a belief in *progress* and an associated fall of the natural man arose to succeed it.

Influenced by Darwin's evolutionary ideas, this belief replaced *static* classical antiquity with the idea that advances in technology would not only make the Noble Savage concept obsolete but lead humanity to inevitable extinction! Evil, though not openly acknowledged, was a component of this destiny. In fact, death would be the means of extinction.

So it is appropriate to evaluate this concept. Where is humanity headed, and what role does evil – and its result of death — play?

In his provocative book *At the End of an Age*[2], historian John Lukacs decries unthinking believers in technology and economic determinism who need to radically rethink – about progress, history, science, the limitations of our knowledge, and our place in the universe.

Frequency and Severity of *War*

Warfare – marked by major conflicts like World War I and World War II – continues to provide undeniable evidence of widespread evil. The threat of worldwide terror by militant Islam seems to be rising unabated. Nuclear warfare looms constantly throughout the world like a Damoclean sword precariously hanging by a thread.

Yet in my lifetime, world population has increased from two to seven billion souls despite these horrible scourges that have needlessly killed millions. So the predicted extinction of humanity has yet to appear, as the birthrate continues to overcome extermination.

Fragmentation — Not Unification — of Interests

One of the remarkable products of technology is readily available instantaneous communication throughout the world. It would seem to enable – if not portend – removal of misunderstandings, distortions, perversions and conflicts originating in social intercourse that have traditionally led even eventually to war.

Likewise, technology has produced means for affordable international travel — thereby bridging cultural barriers and uniting people groups and nations that were traditionally worlds apart and subject to suspicion and isolation.

International activities like exchange programs as well as goodwill services in medicine, music, and the arts to facilitate understanding amongst diverse people groups continue to flourish.

Yet and sadly, division continues to thrive worldwide. The most notable attempt to unite the world and resolve differences – the United Nations – is in shambles today. It is best known for its squabbles, not its solutions. Suspicion reigns supreme. To deny that evil is rampant worldwide would appear absurd.

Public v. Private *Lawlessness*

The rule of law has been a mainstay of civilization – providing a framework or skeleton upon which differences are resolved, rights are established, and injustice is righted. It is obviously dependent on agreed *values* for right and wrong, acceptable and unacceptable, and the ability to establish and enforce those values.

What is the impact of *evil* on the law? How does *evil* manifest itself? Why is *evil* a significant factor in law enforcement?

Evil could be considered to be the fundamental *reason* for law! Were there no evil, law enforcement would virtually be unnecessary. Of course, there would be differences of opinion and behavior based on misunderstandings, but evil as a motivator would be absent.

Although not a recent factor, *public* lawlessness has become increasingly prevalent — at least in the United States. Politicians have emerged as violators par excellence, virtually with impunity. Greed, deception, and fraud abound as evidence of evil. Weaving and wending through an increasingly complex system of laws, they are successful in avoiding any penalty for breaking laws.

In summary, all evidence points to the conclusion that evil continues to abound in the 21st century society with insufficient capability to eradicate it – and engendering death as its final product.

Chapter 6

Modern Masking of Death

"To practice death is to practice freedom. A man who has learned how to die has unlearned how to be a slave." — Michel de Montaigne

Arriving home late from the office on 17 July 1996, I was exhausted. I had already removed my shirt — intending to get to bed early and was in the kitchen, preparing to dish up my nightly ice cream. Suddenly, my wife Phyllis screamed from the bedroom "Get in here – there's been a terrible crash!"

Rushing to our bedroom TV, I watched Chris Wallace (*NBC News*) describing the explosion of TWA Flight 800. Shortly after taking off from JFK International Airport in New York, it had blown apart – instantly killing all 230 people aboard the 747 jumbo jet.

Only 2 minutes later — as I continued watching, the phone rang. It was Gail Chalif (CNN Atlanta) begging me to come to the Washington DC studio as soon as possible. I got dressed right away, and by 10:22 PM, I was being interviewed on the CNN network.

Very shortly, after responding to the first few CNN segments from New York, CNN Senior Vice President Gail Evans called me — right on the set between interviews. "Please agree to stay all night! I'll pay you extra!" she yelled. They promised to register me at the nearby Washington Court Hotel afterward. I had no choice . . .

Believe it or not, I was on camera as shown from that moment until 04:20AM the next morning – a record of six sustaining hours! It is best described as chaotic – both in topic and frequency of interruption for "breaking news."

Typical CNN appearance during six continuous hours that began at 10:22PM – less than 2 hours after TWA Flight 800 exploded on 17 July 1996

During this time, I made one visit to the men's room – trailed by Martha Raddatz (*NPR Pentagon* correspondent) seeking an interview for *NPR Morning Report*. So we stepped into a small nearby room where she interviewed me for 5 minutes before I rushed back on the set to continue responding to new details from New York that kept coming in.

Food was brought to me periodically that I gulped down between interviews.

A limousine finally took me to the hotel at 4:30AM where I literally crashed into bed – having been up 45 minutes short of 24 hours! Scheduled back on camera at 9:00AM, I got little sleep. Having no razor, comb, or change of clothes, I felt like a zombie when I arrived back at CNN.

Meanwhile, CNN had failed to call Phyllis — as promised — to notify her that I would be gone all night. So she had a harried nighttime, worried about where I was. My office called to tell me that they had finally confirmed and notified her where I was – still at CNN.

The day erupted with calls for more interviews – *Time* magazine, *CBS This Morning*, *FOX Morning News*, *CFRB Radio* in Toronto and many others bombarded me all day.

Thus began a series of 170 interviews that I have given concerning TWA800 on many networks – national and international — over a span of several years... more than anyone else in the world. The cause of the crash remains an unresolved controversy today – considered by many to be a conspiracy ranked as significant as that surrounding the JFK assassination.

Yet – of all those 170 interviews about TWA800 — one stands out in my memory above all others...

Increasingly Private Longevity

Whenever Sam Donaldson's name is mentioned, most people recall his loud calling out questions to Presidents — as ABC's White House Correspondent. His own book, *"Hold On, Mr. President!"*[1] calls him "the brashest, most irreverent, most provocative television reporter in Washington."

In November, several months after the TWA800 disaster in July, *ABC Prime Time Live* called my office to schedule a meeting with Sam. A limousine picked me up for a 10:00AM meeting in ABC's Washington studio.

Sam warmly greeted me as I arrived. We were together for two hours, and he was most cordial – quite a contrast to his public image. I found him to be gracious and sincere — but intense. To start, we briefly discussed his new role as co-host with Cokie Roberts on *ABC This Week* that was about to launch.

However, I was surprised by the subject that dominated almost the full two hours we spent together: *death*! Sam focused the entire time on what really happened *physically* to the 230 people who died on TWA800.

He asked me what I thought it would be like to have *experienced* the explosion, how it would have been initially *perceived by passengers*, how it *felt* when the aircraft disintegrated, *when* I thought they must have died, and what *type* of death they suffered.

It was obvious that Sam believed that all the preceding months of news coverage of the disaster had focused on the *physics* of the explosion – whether it was a fuel tank ignition, a bomb in the baggage hold, or a missile that impacted it. Instead, he seemed more interested in the *human* aspect — the agony experienced by 230 men, women, and children prior to dying.

Over and over, we would review activity on the flight — from takeoff through the 12 short minutes that transpired before it disintegrated in a ball of fire, propelling all those bodies out into the air before they plunged downward into the ocean. *What* – Sam wanted to know — did I think *actually* took their life? Was it fire, dismemberment, hypoxia, or drowning?

By this time, the Suffolk County Medical Examiner had already months earlier declared that he believed the 230 were all totally unconscious or dead by the time they hit the water. He had vividly described the presumed bodily impact of the explosion — "It's an extremely violent whiplash... an instant loss of consciousness."

Yet autopsies had shown that quite a few bodies had contained shrapnel, so there was reasonable ambiguity about the actual mechanism of death. In other words, death still remained an intriguing *mystery*.

Sam Donaldson's intense interest in the *process of dying* upsets or challenges a modern trend concerning dying – death being increasingly a *private affair* not shared with anyone.

It was not always so.

Common Deathbed — No Longer *Common*

The recent and continuing increase in human longevity has been matched with a corresponding "privatization" of human death. Not only are people living longer – they seldom die privately amongst their family in anticipation of death.

In the United States until early in the 20th century, most people died under 50 years of age in their own homes in bed due to a short illness. It was common for that person's family to gather in the home around the bed to be present for the death — something expected as normal. The term, "deathbed," was a familiar one.

Two phrases that were frequently heard regarding someone who was dying were *deathbed conversion* and *deathbed confession*. The former referred to a person who had openly disavowed religious faith throughout his or her lifetime finally accepting the reality of God to believe in life after death. The latter often solved hitherto unknown facts or a key role played by a dying person in a crime, love affair, or forgery.

A renowned deathbed confession was made by the famous Israeli singer and songwriter Naomi Sapir who died in 2004 of cancer. Shortly before her death, she wrote to a friend, saying she had used a Basque folk melody as the basis for her renowned 1967 Israeli anthem "Jerusalem of Gold." She had always denied it before. The friend and her family decided to publish the account.

Intellectual Views of Death

There are several types of people who are directly involved in the process of dying and death as a profession. This includes physicians, psychologists, hospice workers, and spiritual advisors working in private settings with people are nearing death.

Committed to assisting those who are about to die, these professional experts have published certain conclusions about the death experience derived from their observations of what most people might consider irrational – lacking available reason or understanding. The most obvious limitation for these experts is their inability to prove the validity of their conclusions, since all such views can never be tested for validity. They are all *preparatory* in nature. They all lack *confirmation* of their efficacy. Yet they all provide a thoughtful perspective on ensuing death.

Sherman B. Nuland, a distinguished surgeon, spoke of the role of hope in his book, *How We Die: Reflections on Life's Final Chapter*[2].

"A promise we can keep and a hope we can give is the certainty that no man or woman will be left to die alone. Of the many ways to die alone, the most comfortless and solitary must surely take place when the knowledge of death's certainty is withheld."

Dr. Nuland — who avowed a "unique relationship" with death — died of prostate cancer at home in 2014.

Marie de Hennezel is a psychologist in a Paris hospital's palliative care unit for the terminally ill who also works with the dying at an AIDS hospice and in their homes. She says in her book, *Intimate Death: How the dying teach us how to live* [3]:

"Our recent experience confirms that the person who can say to someone else 'I am going to die' does not become the victim of death but, rather, the protagonist in his or her own dying."

Another psychologist, Kathleen Dowling Singh, wrote a book as "an ordinary person working with ordinary people dying ordinary deaths." In *The Grace in Dying: How We Are Transformed Spiritually As We Die* [4], she says:

"In the face of death, we cower. This is said not with judgment but with utmost compassion. Even after having anticipated in the mystery of death so many times and have come to know its unquestionable beauty and majesty, I still cower at the thought of 'my turn.'"

Privacy of Personal Death

As transportation and technology have created an increasingly mobile and complex modern society, death has been almost surgically removed from public consciousness on a *personal* basis. Most announcements of death – particularly of the famous – come as a shock for the public, followed by profuse praise of their contributions to life. They have died quite privately and quietly out of sight.

However in contrast, death is increasingly pursued and reported on a global or *impersonal* basis. AIDS, global warming, immunity to prescribed pharmaceuticals, airline disasters, and random terrorism are examples of headlining impersonal death.

The "fatality density" associated early with virtually inexplicable circumstances of an airline crash will frequently result in an overwhelming flood of TV and news coverage — filled with speculation, rumors, and chatter. Typically, it persisted for days after the 2015 suicide dive of the Germanwings Flight 9525 into the Alps. And I was among those pressured to provide such interviews.

Ironically, though 150 were instantly killed in that tragedy, on that same day in the United States, over 350 people died just as suddenly, tragically and unexpectedly in *auto traffic accidents* with little public awareness or concern. Have you ever wondered why the public is more interested in *massive* deaths than in *singular* deaths?

Ambiguity and Confusion of Death's *Definition*

Occasionally it is said in a joking manner, "You are worth more dead than alive." However, that may be more truth than fantasy – depending on the *definition* of death.

The introduction of human organ harvesting has obviously resulted in stricter definition of *when* death has occurred – because organ transplants cannot occur until death of the donor has been certified. The magnanimous generosity implied when you check the "organ donor" box on your application for a driver's license carries with it some potential complications.

The 1968 Uniform Anatomical Gift Act (UAGA) was intended to harmonize state laws covering organ donations for the purpose of transplantation – since there are differences between them. This Act also governs the making of anatomical gifts of one's cadaver to be dissected in the study of medicine. It prescribes the forms by which such gifts can be made.

UAGA also provides that — in the absence of such a document — a surviving spouse, or if there is no spouse, a list of specific relatives in order of preference, can make the gift. It also seeks to limit the liability of health care providers who act on good faith representations that a deceased patient *meant* to make an anatomical gift.

One concern here is that death's *definition* is not totally patient-focused and instead accommodates the needs of the medical profession and organ transplants. So organs can be harvested from an irreversibly brain-dead person, thereby legally dead, even though circulatory and respiratory functions are still ongoing and viable.

Organ transplantation thereby puts apparently unstoppable pressure on the traditional *legal* definition of death. And that earlier joke about your value may become a reality!

Shifting from Personal to Impersonal Death

The focus or interest in dying is changing for many today. Medical science continues to pursue health issues, of course. And specific types of death-causing diseases like cancer, ebola, and diabetes are researched with intensity.

However, concern about death is beginning to be addressed increasingly from a perspective more related to the *collective impact* of personal and individual actions than to *individual* human mortality itself. Concern about death at the personal level seems disguised or hidden – seldom openly acknowledged.

Whereas plagues and sweeping epidemics historically drew apprehension and fear of death – as did the worldwide influenza following World War I, there is an entirely new lure for global concern – *environmentalism*. This belief is a cult that condemns humanity for its insensitivity about how it selfishly utilizes and abuses natural resources. *Everyone* is seemingly guilty of polluting, desecrating, and ruining planet Earth — our home.

This resulting universal "guilt trip" closely resembles the Biblical declaration that "all have sinned and fall short of the glory of God."

Before analyzing this widespread cult, I want to acknowledge the unnecessary and abusive destruction of nature that I myself have personally witnessed. As a guest of the Czech government immediately following the collapse of Soviet occupation of Bohemia, I visited and photographed horrible scenes of the physical effects of mining and burning that region's soft brown coal on forests and water table. The health effects, as well, were devastating. Thousands of children pitifully died of resulting emphysema.

So the results of technological, political, and financial interests are often disastrous. Effort to challenge and control those forces in order to retain healthy living is both admirable and essential. However, like many other belief systems, environmentalism has emerged as a force to be examined and challenged.

The worldwide environmental network is focused on — and virtually promotes — *worship* of the earth. Its conception occurred in April 1970 when Earth Day launched the environmental movement. The Earth Day Network (EDN) works with over 22,000 partners in 192 countries to broaden, diversify and mobilize this crusade. More than 1 billion people now participate in Earth Day activities each year, making it the largest civic observance in the world.

EDN works to broaden the definition of "environment" to include all issues that affect our health, our communities and our environment — such as greening deteriorated schools, creating green jobs and investment, and promoting activism to stop air and water pollution. Yet *delaying death* – the unstated objective of a habitable earth — is never mentioned.

Amazingly, the environmental movement has practically become a *theological* crusade – intended to assume responsibility for protecting Earth as a part of a *covenant with God*! As an example, the Religious Action Center of Reform Judaism (RAC) has created a guide for connecting its Shabbat observance with Earth Day. It has also created Earth Day Shabbat – the weekend preceding Earth Day – as a means for *urging clergy leaders of all faiths* to bring messages about the environment into their congregations.

Why has this alarming concern for the Earth suddenly appeared? More interesting is why the focus is on the *Earth*, rather than on *human mortality* – the obvious result of a degenerating environment. Could it be that *death* can no longer be discussed in the public arena? Must death's role – as a motivator of accountability – be hidden?

Consider for a moment the possibility of successfully "freezing" any more change or degradation of the earth – no more global warming, no more air or water pollution. Were that to occur, what would happen to humanity? Would they still *die*? Yes. Would the *mechanisms of death* remain the same as they are today? Yes.

Why then not address the *inevitability* of death — as well as options for dealing with it?

The theological twinge of environmentalism raises another possible corollary to its misplaced emphasis on a *temporary* Earth instead of the *finality* of death. Paul, the famous apostle, in a letter to Roman residents decried those who "worshiped and served *created things* rather the *Creator*."

Technology's *Disguises* of Death

There is no doubt that technology has produced many blessings for humanity. It is easy to recognize technology in society. It consists of devices, knowledge, and skills that can be used to control and utilize physical, social, and biological phenomena.

Travel, communication, computation, and health care (medicines, life support, anesthesiology, diagnostics, and artificial limbs) all testify to technological expansion and enjoyment of life.

But the relationship between technology and *death* needs examination. There are two sides of that ledger. Only one is advantageous.

Advances in medicine can prolong life, but they can also make it more difficult for doctors to know when a patient has truly died. As medical technology becomes more advanced, it also becomes increasingly challenging for doctors to discern the line between life and death. It also complicates how doctors might approach the end of life.

A great advance of technology – the PET (positron emission tomography) scan — is an imaging test that uses a radioactive substance called a tracer to look for disease in the body. It shows how organs and tissues are working – actually detect cellular activity in tissues. This is different than magnetic resonance imaging (MRI) and CT (computed tomography), which show the structure of and blood flow to and from organs.

Families and clinicians could wonder if all patients in a coma should have a PET scan, or whether patients whose scans show residual brain activity should be kept alive longer than he or she would have been otherwise. But what can never be determined is whether a recovered patient will regain the kind of life they would find meaningful. Being *alive* is not enough.

In a similar vein, an emerging resuscitation technology called extracorporeal membrane oxygenation, or ECMO, is complicating not how doctors pinpoint a *time* of death, but how and when we admit that death is *inevitable*. Death, death, death... an elusive finality.

Before crediting technology too much for *precluding* death, we must not overlook another of its results – *massive destruction*. Technology has allowed each new generation the ability to create increasingly destructive means of warfare like the use of laser-guided missiles, unmanned drones, biological warfare, and nuclear weapons.

Governments as well as radical individuals use biochemical and nuclear threats to confront social, economic, and political vested interests. Violent actions increasingly result in targeted mass destruction and deaths such as occurred on 9/11.

Denial of Death

Ernest Becker, a Jewish-American cultural anthropologist and writer, won a Pulitzer Prize in 1974 for his book, *The Denial of Death* [5].

The theme of his book is that man is the only creature who must pass a lifetime with the fear of death haunting even his most sun-filled days. Becker held that man's innate fear of death is a principal source of his activity. He said that one of the great rediscoveries of modern thought is that of all things that move man, one of the principal ones is his *terror of death*.

Many believe that Becker's greatest achievement was the creation of the "science of evil" – wherein he believed that individual character is essentially formed around the process of denying one's own mortality. This denial is a necessary component of functioning in the world. He held that much of the evil in the world was a consequence of the need to deny death.

Fear of the *Uncontrollable*

It is human nature to experience fear – to possibly worry about the variety of risks discussed in Chapter 2. However, to have *constant* fear of uncontrollable circumstances can lead to anxiety and concern that far exceeds normal, cautious behavior. And that is irrational. Such anxiety can end up controlling a person.

At the outset of a fearful situation, it is wise to ascertain whether it can be controlled – either by your own action or by enlisting help of others. Police, fire, and emergency medical personnel are examples of available sources of assistance if the predicament exceeds your personal capability. But they are not always available . . .

Should it be obvious that conditions are beyond your ability to control, rational thought is to center your mind on either (a) prevention or (b) taking an active role in the inevitable outcome. Given that death is inescapable and cannot be prevented by any amount of worrying, fretting, or concern, it is wise to contemplate it rationally and take *systematic action* that precludes fear.

Fertile Ground for Political *Manipulation*

Death lurks as an agent virtually undetected by our senses. While we are made aware of death by a variety of factors – departure of family members and friends, news media accounts of accidents, crime or warfare, as well as personal health issues, we try to push it out of daily awareness or concern. This ongoing but subconscious apprehension of death's reality creates a productive arena for exploitation by advertisers, moralizers, and other persuasive parties.

Perhaps the most recognized persuaders are politicians whose very existence is dependent on recruiting, aligning, and mobilizing large segments of the public to elect and support them. A dark side of political life could be described as *manipulative* – as contrasted with *informative*. Political manipulation is the latent management of citizens — influencing their societal consciousness, values and behavior — by politicians with the purpose of forcing them to operate contrary to their own self-interest.

Death, for the politician, is an operative and leveraging factor that subtly appears in such diverse fields as war (DoD), safety (OSHA), health (FDA), travel (FAA), housing (FHA), and retirement (HHA).

Stripping Off Death's *Mask*

The key to reality lies in recognizing, acknowledging, and living with one's ultimate destiny – *death*. Because it could easily be viewed as an anathema to be avoided or denied, death is deliberately masked in much of society.

"Oh, don't even *mention* it!" many will say. Others may remark that pondering the subject *frightens* them. But why? What is it that motivates censorship of life's *finality*? Could it be that inadequate thought has been devoted to its reality?

When something as real as death is forced to be artificially banished from discussion or serious thought, life can become meaningless — without a realistic goal. Openly acknowledging death need not put a damper on life any more than acknowledging the boundaries of a football field limit enjoyment of that sport. Both admissions are real.

And being realistic about death is not necessarily a religious idea, though often thought so. The agnostic George Santayana once said: "Nothing you can lose by dying is half as precious as the readiness to die, which is man's charter of nobility."

Incomprehensible Via Specious Shift

Death in modern times has become "off limits" for normal, rational conversation in the home or on the street. One possible reason for this is that it has become necessary to use code words – or even humor — to disguise it. Death seems just too strong, realistic, or severe.

Comedian Woody Allen *joked*, "I'm not afraid of death; I just don't want to be there when it happens." Much earlier, the last words of French

Renaissance humanist François Rabelais spoke of death as a *journey*, "I go to seek a Great Perhaps."

What if *banking, engineering,* or *education* were described in such obtuse terms? Somehow, though death is a universal event for every single human being, it remains in an independent status – and described in poetic, fanciful, ethereal terms well outside common discourse. Why? Primarily because there is no *post-event* experience to share, discuss or analyze.

But the result of this void of acknowledgement turns out to be shock and morbid interest that is generated as death's reality explodes. "Breaking news" or dramatic headlines interrupt everything... Time stands still.

The conclusion that German philosopher Georg Wilhelm Friedrich Hegel reached about death 150 years ago might help to overcome the prevalent and widespread unrealism that exists.

Hegel's principal achievement was his development of *absolute idealism* as a means to integrate the notions of mind, nature, subject, object, psychology, the state, history, art, religion and philosophy. How did Hegel view *death*?

"If I take death into my life, acknowledge it, and face it squarely, I will free myself from the anxiety of death and the pettiness of life and only then will I be free to become myself."

Fallacy of "*Saved*" Lives

Another disguise of death is heard virtually every day. It is the phrase, "X number of lives have been *saved* this year by _____. (Fill in the blank.)

The widespread fallacy of "saving lives" by blood transfusions, organ transplants, seatbelt use, hardhats, smoke alarms, airbags, bicycler helmets, or quitting use of tobacco is part of this charade. Obviously, those lives are not *saved* — only the *mode, conditions,* and *timing of death* are revised.

The implication is that all those people would have *died*, were it not for bodily protective equipment, medical intervention, or warning. The intent is admirable, of course, but inevitable death was neither acknowledged, mentioned nor prevented.

What does it mean to "save" a life? Can life be *saved*? Is it not more accurate to say that *life* was extended or *death* was delayed?

Immortality Via Interchangeable Body Parts

How awesome and magnanimous it is when someone volunteers to allow surgical removal of their kidney, lung, or cornea to be transplanted in another person – especially when they themselves are still living.

Though increasingly happening today, it was only in 1954 that the first human organ transplant occurred – the kidney of an identical twin. The recipient lived another eight years, and the surgeon who performed the operation won a Nobel Prize for his skill.

The practice of transplanting organs has expanded and improved considerably in recent years – to the point that the shortage of *donors* is now the key constraint to organ transplants. Congress has passed numerous legislative actions supporting organ donation. Several Surgeon Generals have personally appealed for more organ donors, and Medicare now pays for donor transplants. Despite this strong support, waiting lists for organ transplants have continued to grow until well over 100,000 Americans are now waiting for a donor organ.

Certainly life is to be treasured – even extended beyond death if possible. But the larger issue, obviously, is whether life is all there is to reality. Obtaining replacement body parts or functions must be weighed against some baseline, value, or benefit.

At the end of World War II, I worked as a teenager in a large auto parts warehouse – to pay for my college tuition. Most of my co-workers were recently-discharged veterans – a rough and tough group who often recited competitive and gory – even frightening — details of their combat experiences. It resembled a time of "Can you top this?" I listened – with awe.

One of those hardened vets – Arnold – was taciturn and tight-lipped, seldom joining in the frequent roustabout boasting of how close each of them had come to being killed in combat. He patiently tolerated their stories without contributing. However, during one bragging session, Arnold apparently had heard all he could tolerate.

Whirling around with fire in his eyes, he screamed, "What do you want to do – live *forever*?" It was a show-stopper, believe me.

Somehow, Arnold had concluded that all these accounts of "dancing with death" were meaningless unless *dying* could ultimately be avoided. Why were these veterans *proud* of having escaped death? What if immortality is attainable – even a *possibility*?

Chapter 7

Death: Finality or Change of State?

"Dum Spiro, spero- As long as I breathe, I hope." — Marcus Tullius Cicero

"And that concludes my address *auf Deutsch*."

As the first of four American guest speakers, I had just attempted to be gracious by speaking a few introductory sentences in German to begin my presentation. Having studied only one year of German in college 25 years earlier, it was obviously intended to be humorous rather than serious. And I expected that the 650 German executives and academicians in the impressive Koelnmesse would see it that way.

However, I was shocked when there was not one smile in the huge male audience. Every one of them was stoic – almost seeming to glower at me!

The occasion was the *1972 Institut fur Unfallforschung Colloquium* being held in the huge exhibition center beside the Rhine River in Cologne. We Americans had been invited to share our involvement in NASA and Apollo missions to the Moon. The two-day theme was on applying lessons learned — of systematically managing the high risks of space travel — to the West German economy.

As I finished speaking, an unexpected loud and extended rumble erupted throughout the auditorium. It was my first exposure to German applause – everyone loudly rapping their knuckles on desks or tables instead of clapping their hands.

I descended the dais to be seated beside one of my German hosts – still a bit confused by the impassive reaction to what I intended as a bit of humor before launching into serious topics. Softly I whispered, "Herr Schnadt, we often begin serious talks or discussion in the US with a bit of humor. So I apologize for my apparent gaffe . . ."

He whirled in his seat, turned to face me, and emphatically said, "Germans do *not* mix work with humor! When I leave for work in the morning, I say to my wife, 'I go to my *duty*!'"

Later that evening, our team was hosted by a 3-hour night dinner cruise aboard the double-decked motorship "Rheingold" on the Rhine. In our honor, the Cologne Burgermeister (mayor) ordered colored floodlights to be turned on the long row of historic medieval burgher's houses facing the river. The sight was most impressive as we slowly cruised by! Famous local Koelsch beer flowed freely during the cruise, as the sedate-in-the-daytime German professionals whom Herr Schnadt had earlier described to me were radically converted into hilarious-in-the-nighttime characters.

The person who had invited us to the Colloquium was only a few months younger than I. Professor-Doctor-Director Peter C. Compes (yes, Germans address each other by all three titles simultaneously) had survived WWII and the 95% near-destruction of Cologne by Allied bombing. His father had been an admiral involved during the war in Nazi Abwehr intelligence.

I was a guest in the Compes home in Cologne for nearly a week, surrounded by his gracious wife Marieanna and their five charming children. Peter was active as a lay leader in the local Roman Catholic parish, so I attended Mass with him as he participated in the liturgy. I even spent many hours *nude* with him in his elaborate basement sauna – a new experience for me.

With Peter Compes (left) upon my departure from his Cologne home on 8 May 1972 at the conclusion of the Colloquium

A few months later, I was able to return Peter's hospitality by hosting him in our California home where we swam together in our pool, and he spent time with my wife and children.

Yet throughout the years that Peter and I worked together, I could never fully erase wonderment about the twist of history that converted him from my *enemy* to my *friend*. War creates crazy dualities or dichotomies. Obviously, Peter and I never *knew* each other as enemies – but we *were* nonetheless during World War II!

During World War II, both Peter and I could well have been fighting in the Battle of the Bulge – as enemies. Obviously, it would have been my objective to *kill* him! What happened to *reverse* that status? Something amazing. Politicians, statesmen, and military commanders suddenly and mutually agreed one day to stop fighting each other by simply signing a piece of paper in a ceremony. *Killing ceased.* At that moment, millions of people were transformed from enemies to friends – with no awareness on their part!

If you were on Mars or some distant place – viewing this behavior between otherwise normal human beings, would you not be amazed and confused?

Schizophrenia is a severe brain disorder in which people interpret reality abnormally. When *nations* suffer from this "long-term mental disorder of a type involving a breakdown in the relation between thought, emotion, and behavior, leading to faulty perception, inappropriate actions and feelings, withdrawal from reality and personal relationships into fantasy and delusion, and a sense of mental fragmentation," it is called *war*.

Schizophrenia: Total Cessation v. Metamorphosis

Humanity also suffers another prevalent form of schizophrenia — about *death*. Mutually contrary views are held in suspension: whether death is the *total cessation* of being or only a *metamorphosis* of being — like a cocoon awaiting transformation into a butterfly.

Studs Terkel, a prolific Pulitzer Prize-winning author, wrote a book at age 88 that addressed what he called *"the ultimate human experience: death and the possibility of life afterward."* He recorded 63 oral interviews of a wide variety of people – aged, young, celebrated and ordinary, brothers, religious and atheists, strangers, friends, and acquaintances. His book — *Will the Circle Be Unbroken? – Reflections on Death, Rebirth, and Hunger for a Faith*[1] — provides a remarkable spectrum of thought that confirms schizophrenia concerning death.

Linkage of humans with other mammals has also contributed to death schizophrenia. On one hand, some folks fully equate *animals* with *people* — being even more concerned about the death of an endangered species fetus than a partially-born human. On the other hand, following massive public disasters – earthquakes, tsunamis, tornados — causing thousands of deaths, almost universal desire arises to believe that there is *reward* after supreme sacrifice, *comfort* after innocent ravaging and *life* after death. In other words, instantaneous cessation of being seems too severe to be acceptable.

Whether it is possible to attribute schizophrenia – an individual brain disorder — to humanity at large may be debated. However, there is little doubt that humanity appears to be *bi-polar* – holding two diametrically opposed views – concerning death.

On 9 April 1951, General Douglas MacArthur – having been summarily relieved of duty in Korea by President Truman for insubordination and recalled to the US – addressed a joint session of the Congress. At the conclusion of his emotional "farewell address," MacArthur said:

> *I still remember the refrain of one of the most popular barrack ballads of that day which proclaimed most proudly that* '**old**

soldiers never die, they just fade away' . . . *I now close my military career and just fade away.*

His physical death — almost another 13 years later — was certain and factual. On the other hand, his "fading away" illustrates the idea of death schizophrenia. Was MacArthur's death *final* or *metaphoric*?

Siren of Scientism

Science has contributed so much to the human race. Its descendant prodigy – technology – continues to bedazzle us with incredible labor-saving benefits. That reputation, among other factors, has increasingly encouraged and popularized a mindset known as "scientism." Defined as "the tendency to reduce all reality and experience to mathematical descriptions of physical and chemical phenomena," scientism has a charm that seems to offer an explanation of "all that is or ever will be" in nice, clean, understandable terms.

Since death – especially the *results* of death on reality – lies outside any experimental examination, there is no way that it can be subjected to scientific examination or technical elitist language that attempts to reduce all reality to mathematical description. Beyond that obvious limitation, death defies *any* scientific examination, test, or analysis.

Certainly all of us would like to have an "explanation for everything." That's the tantalizing enticement of scientism. Imagine how exciting it would be to have an "app" on your hand-held device that not only revealed all your vital signs — exactly how, when, where, and why your death will occur – but also precisely what will occur to you *after* you die!

Though it is never openly identified or acknowledged, scientism is always in the shadows – silently promising to someday rationally describe the *meaning* of death in scientific terms. Behavioral scientist Isador Chein once described scientism as "arrogant, pompous, and exclusionary."

Post-Death Influences

The abrupt cessation of human life has often caused speculation – even hope – that some undetectable aspect or influence of a person continues to live on beyond physical death. It is sometimes expressed in the form of a question, "Do we have an eternal *spirit* that is not subject to dying?" Of course, there is no *physical* means for measuring or determining the existence of such an essence. Ghosts, phantoms, zombies are frequent topics for entertainment mysteries, but they do not represent common conviction of reality.

Preparation of written wills is one means for attempting to exercise influence after death. Distribution of accrued wealth – provided someone is available, willing, and able to carry out expressed desires – enables the deceased to wield power and authority beyond the grave. But wills are not infallible sources of posterity. They can be ignored or deliberately violated.

Architecture provides physical materiality of influence exercised by departed individuals. The pyramids of Egypt, Taj Mahal in India, and Frank Lloyd Wright homes all provide evidence of those who have previously lived. There is a continuity . . . an ongoingness. . . a descendance that testifies to previous existence.

Written documents – books, papers, correspondence – are an avenue for remembrance of authors beyond death. Authorship enables influence on all readers but is generally limited to those who understand the milieu and language employed in writing.

Politicians, regardless of their momentary notoriety, have a very short *half-life* of significance. In the United States, the persuasive authority and political significance of potential Presidents Spiro Agnew, Gary Hart, Nelson Rockefeller, Ed Muskie, Dan Quayle, and Michael Dukakis faded far faster than it rose to public awareness. Despite all the news coverage they received for a short time, little influence of their political contribution remains. It has disappeared much like a wisp of smoke in a breeze.

Expiration at Burial

Cemeteries are an interesting phenomenon around the world. They frequently represent very expensive real estate located in the middle of the busiest section of major cities. Traffic is diverted around them. Grounds surrounding gravesites are beautified and constantly maintained. Silence and respect is anticipated. At varied holidays, cemeteries often become places of beauty as graves are decorated with flowers, flags and other memorabilia.

In the United States, Decoration Day was established at the end of the Civil War as a time set aside to decorate graves of the war dead with flowers. Following World War II, it was re-named Memorial Day – ultimately being designated in 1967 as a three-day United States federal holiday to be observed on the last Monday of May to honor men and women who died while in the military service.

Arlington National Cemetery, adjoining Washington, DC, now holds over 400,000 graves dating back to the Civil War. The adjoining Navy Annex — that for many years housed the headquarters of the US Marine Corps — was demolished in 2014 to provide additional burial properties anticipated to be needed for veterans of World War II, Korea, Vietnam, Iraq, and Afghanistan.

Invited to lecture at the Chinese Academy of Sciences in Beijing in 1981, I toured the famed Thirteen Tombs of the Ming Dynasty 25 miles north of Beijing. The long entrance to the imperial necropolises is known as the Sacred Way or Divine Road which means *the road leading to heaven*. Each Emperor, known as the Son of the Heaven, who came from Heaven to his country through the Sacred Way, is also believed to have deservedly returned to Heaven through this road.

The Sacred Way is lined with 24 stone animal statues which are important decorations of the mausoleum — lion, camel, elephant, xiezhi (a mythological unicorn), qilin (one of the four "divine" animals, the other three are dragon, phoenix and tortoise), and horse. There are 4 of each of these animals: two standing and two squatting with different meanings. Lion symbolizes awesome solemnity because of their ferocity. Camel and elephant are meant to suggest the vastness of the territory controlled by the court, because they are dependable transport in desert and tropics. Xiezhi was put there to keep evil spirits away, because it was believed to possess the sixth sense to tell right and wrong. If two men fight, a xiezhi would gore the wicked one. Qilin, an auspicious symbol, was placed on two sides.

Astride one of the horses on the Sacred Way on 4 January 1981. As the emperor's mount, the horse is absolutely indispensable. Legend says that all these animals are supposed to change guard at midnight.

This ancient and elaborate memorial testifies to the centuries-long belief or hope that there is life beyond the grave. Cynics, of course, mock all such investment as foolishness – nothing more than superstition or wishful thinking.

Death as "End of Something"

Part of the ongoing irresolution about what occurs following human death is due to widespread use of the term "death" to describe the *end of something*.

Engines *die*. *Dead* batteries are common. We hear of "death of a vision" or the "dying days" of blacksmiths, the horse-and-buggy, service stations, and passenger trains.

Packard, LaSalle, Nash, Essex, Oldsmobile, Hudson, Plymouth, Studebaker, Terraplane, Willys and DeSoto automobiles have all *died* — being built no more. Famous piston-engine passenger airliners like Boeing *Stratoliner* and Lockheed *Constellation* had fairly short commercial lives and *passed away* to be soon replaced by jet-powered De Havilland *Comet*, Boeing

707, Douglas *DC-8*, Lockheed *L-1011 Tri-Star*, Sud Aviation *Caravelle*, and Tupolev *Tu-104* – all of which now lie in aircraft *graveyards* after their remarkable role in popularizing and multiplying air travel worldwide.

And similar death analogies apply to so many other fields beside transportation – like banking, medicine, tourism, architecture, and law. Why does this happen? What makes death such an applicable metaphor for termination?

Popular Applications to the *Inanimate*

Beyond death being a widely-used image of *termination*, it is also applied to things, situations, and objectives that *never had life* – at no time having lived or existed in a living state.

Consider music for a moment. What is music anyhow? It can be defined as "the art of arranging tones or sounds in succession, in combination, and in temporal relationships to produce a continuous, unified and evocative composition through melody, harmony, rhythm, and timbre." Being *inanimate,* it consists of neither life nor death. Yet it vividly portrays both! There is *lively* music, and there is music marking *death* – like funeral dirges or a requiem — a musical composition in honor of the dead. But can music itself *die*?

Widely used *expressions* also illustrate this attribution to the *inanimate*: dead as a doornail, kick the bucket, dead last, dead of night, dead end, deadman switch, dead of winter, dead to rights – a lengthy vocabulary that uses death descriptively to connote the end of something.

More than likely, applying the term "death" to terminal, completed, final but not living phenomena speaks to the impossibility of influencing the outcome – the futility of opposition. And even more importantly, it reveals the power of death to determine, affect and sway human thinking on a large scale.

Endless Human Life?

The idea of living forever – as an objective or desire – has likely been entertained by many at some time in their life. Remember Arnold's fiery retort in Chapter 6 "What do you want to do – live *forever*?" The unspoken answer is "Yes!"

In other words, *ceasing to exist* is not a natural inclination. In fact, it may be almost impossible to imagine. Would all that has occurred to you in your lifetime suddenly and simply disappear – like feathers in the wind? Was there any meaning or purpose to having experienced everything that happened to you or that you accomplished on earth?

Apparently Secular Humanists conclude there was no purpose in being alive. They come from "nothingness" and return to "nothingness" with purposeless life simply sandwiched in between.

Admittedly, all our lives seem to be a series of random events and conditions over which we seem to exercise little control – starting with our birth. Yet, in contrast, we can perceive a meaningful pattern of happenings that define our biography – our existence, our involvement in humanity, our interactive participation in human history. We have witnessed and can recall former times – "the days that are no more," as Tennyson said.

Yet we are sufficiently aware that we may well be forgotten after we die – because we forget others ourselves! That's why some erect some physical means of being remembered – like the Taj Mahal or the Egyptian pyramids. All recent American Presidents have established libraries to house memorabilia – ostensibly to share items of historical significance that occurred during their time in office but also to retain and support perpetuity of their lives. They, like most of us, want to believe that we all have been significant from the moment of birth – and that our significance has no identifiable end.

Dogs, Frogs, and *People*?

If life is defined as "an organismic state characterized by capacity for metabolism, growth, reaction to stimuli, and reproduction," then dogs, frogs and people share *life* together. Yet these three types of living beings can readily be differentiated by size, shape, appearance, intelligence, and length of life. Another differentiating aspect of these three is *volition* – the ability to make choices among options to reach a conclusion. This ability utilizes and applies *reason* to situations.

A larger question is whether life is a *homogeneous* state for everything that possesses it. For example, flowers, trees, grass have *vegetative* life. And they perpetuate that life without assistance. How is that type of life differentiated from *organismic* life that involves an identifiable input-output, reproductive and communicative process?

The delicate relationship between life and death – and to what extent they influence each other – clouds the possibility of accurately defining death itself. Even though the dictionary defines death as "permanent cessation of all vital functions," does that conclusion resolve whether death is the *absolute end* of life or the *boundary of a new state* for life?

Questions can be provocative and insightful — but do not necessarily produce or even lead to desired answers. Death remains elusive.

The "Wannabe" Syndrome

Many of us have a common reaction when we see another person doing something out of the ordinary – particularly in uncertain situations. We instantly and subconsciously feel a need to try it out ourselves — even if what that person is doing is absolutely worthless. For example, if we observe someone suddenly depart a crowd in which we all have been impatiently waiting – headed for an apparent escape or better route, we may instantly follow them with no evidence of success.

This tendency illustrates a "wannabe" – a person who desires to be, or be like, someone or something else. Other similar behavior has been called the herd instinct, group-think, copycat or devotee of "fads." Collectively, this group of behavioral symptoms even creates what is known as "style," and Madison Avenue thrives on it.

Philosophers Søren Kierkegaard and Friedrich Nietzsche were among the first to criticize what they referred to as "the crowd" (Kierkegaard) and the "herd instinct" (Nietzsche) in human society. Modern psychological and economic research has identified herd behavior in humans to explain the phenomena of large numbers of people acting in the same way at the same time.

Obviously, this characteristic offsets or counters *independence*. From childhood, my father continually urged me, "Don't be afraid to be different!" It was wisdom for which I have always been most grateful.

The "wannabe syndrome" plays a major role in attempting to resolve the post-death dilemma. First, comfort is generated when "everyone thinks the same way" about it. Second, the obvious lack of scientific proof fades to being no longer significant, as long as there is widespread agreement.

Hope Breathes Eternal

What is it to *die*? For some, death is our body's expiration date, for others it is an absolute point where the soul leaves the body. *Life after death* does not die easily.

For a group of scientists in Scottsdale, Arizona, death is merely an *arbitrary natural accident*, an engineering problem we have yet to find a solution for. The Alcor Life Extension Foundation (ALEF) is the world's leading provider of cryonics, the practice of using ultra-cold temperatures to preserve humans until such a time when medicine is advanced enough to restore good health.

The renowned baseball star Ted Williams, upon his death in 2002, was decapitated by surgeons at ALEF where his body is suspended in liquid nitrogen. Williams' body was separated from his head in a procedure called neuroseparation. His head and body were preserved separately. The head is stored in a steel can filled with liquid nitrogen. It has been shaved and drilled with holes. Williams' body stands upright in a 9-foot tall cylindrical steel tank, also filled with liquid nitrogen.

Optimism over the possibility that it may someday be possible to bring humans back to life has been boosted by emerging technologies, in particular *nanotechnology* and its potential at a molecular level to repair and regenerate cells and tissues in the not-so-distant future.

ALEF believes that many problems relating to organs such as the heart and lungs which we currently consider fatal will one day be reversible using nanomedicine, a prediction which has led most of its members to choose just to have their head preserved (called neuropatients).

Dr. Mark Morrison, chief executive of the Institute of Nanotechnology in Glasgow, says: "I can see cryonics having a big effect on stem cells and very simple organs, but I think restoring function to organs as complex as brains is way off into the future." He added that the challenge lies in preserving the hundreds of billions of neurons in the brain and the interconnections between all of these which create our personality.

The greatest drawback to cryonics as a universal answer to death, of course, is that it requires an intact body at death. This would rule out those dying in accidents or disasters where their disintegrated bodies are identified only by their teeth.

Beyond this rather extreme physiological approach, of course, there has been throughout history strong belief that there will be future resurrection and transformation of human bodies no longer subject to degeneration.

Impetus of Collective Death

One of the obvious results of massive accidental death – "fatality density" discussed in Chapter 6 – is not only to generate extensive news coverage but to also demand changes that will eliminate or reduce the possibility of the tragedy from occurring again.

While individual deaths tend to be accepted as normal and expected occurrences, grouped collective deaths are thought to be *preventable* – at least regarding their timing for those killed. The public views them as tragedies, catastrophes, or disasters due to failed policies, overlooked governmental responsibility, or short-sighted community concern. War, natural disaster, and plague are examples of demand for public intervention and change.

Abraham Lincoln's call at Gettysburg in 1863 provides an example of the demand for change following massive death:

> *It is for us the living, rather, to be dedicated here to the unfinished work which they who fought here have thus far so nobly advanced. It is rather for us to be here dedicated to the great task remaining before us—that from these honored dead we take increased devotion to that cause for which they gave the last full measure of devotion—**that we here highly resolve that these dead shall not have died in vain...***

The unfinished work that had already required many thousands of deaths was at stake. Dying in vain would have meant that rebellion would have continued unabated. Lincoln's resolve, per se, provided adequate impetus for massive combat deaths to have meaning in another two years. Sadly, warfare has continued worldwide unabated ever since – suggesting that millions have died in vain.

Can't Have It Both Ways

Whether death is the *absolute end* of life or simply the *boundary of a new state* for ongoing life, both remain as highly-contrasted alternatives.

Unfortunately, these two opposing beliefs cannot co-exist. Neither is subject to scientific proof.

So what is the answer to this dilemma? Is it a lottery – casting of lots – Russian roulette? Or something else?

Chapter 8

One Answer From Antiquity

> *"Tarde quae credita laedunt credimus (We are slow to believe what hurts when believed)"* — Ovid

I needed an *answer* right away — desperately!

Only two days earlier, I had arrived in Havana, Cuba – with two other musicians – to participate in 10 days of open-air gatherings throughout Cuba arranged and conducted by a dear friend, Waldo Nicodemus.

The three of us – Marv Strum, Val Johnson and I — were from Seattle. Marv had just purchased a brand-new red 1956 Chevrolet, famous for its powerful V-8 engine, in which we likely set a remarkable speed record by driving *non-stop* from Seattle in the Northwest to Key West, Florida – the southeastern tip of the US.

There were no Interstate highways in 1956. Yet we covered *3600 miles in 72 hours* – averaging 50 miles per hour. And that included stopping long enough for a lube and oil change!

Boarding the SS "City of Havana" in Key West, we sailed to Havana – a little over 100 miles away. Waldo was there to greet us. He and his family lived in San Francisco de Paula – a suburb of Havana. Just a day before we arrived, its sewage system had somehow failed – a major inconvenience for the Nicodemus family as they graciously hosted us.

Sadly, it was only two days later that I suffered an attack identical to what President Carter inelegantly described at a 1979 Mexico City luncheon hosted by Mexican President Portillo:

"In the midst of the Folklorico performance, I discovered that I was afflicted with Montezuma's Revenge." (This *faux pas* ricocheted around the world, forever embarrassing Carter.)

The stomach and intestinal infection known as Montezuma's Revenge is the most common illness affecting travelers. Sometimes it can be severe enough to require medical intervention. The destinations of highest risk are countries in Latin America.

The primary source of infection is ingestion of fecally-contaminated food or water. Evidently our sickness was a result of water we were drinking in San Francisco de Paula. Travelers often suffer this illness from eating and drinking foods and beverages that have no adverse effects on local residents, so Waldo and his family were unaffected.

Meanwhile, all three of us were soon afflicted. I was determined to find an *answer* because there was no way we could even start our tour with this malady.

The first remedy that came to mind was to find some Pepto-Bismal – over-the-counter medicine that I recalled my nurse wife keeping in our home as a general cure-all. It claims to remedy heartburn, indigestion, upset stomach, diarrhea, and other temporary discomforts of the stomach and gastrointestinal tract.

As we walked toward an open-air restaurant where Waldo was escorting us for lunch, we passed what I assumed to be a drugstore where I thought Pepto-Bismal might be available. So as we got ready to order food, I excused myself to return there — hoping to purchase it. Not speaking Spanish, I was finally able to write "Pepto-Bismal" on a piece of paper. The clerk brought out a large bottle (at least a half-quart) that I gladly purchased.

As I re-joined the group at lunch, I proudly held up the hopeful "answer" for all of us — except Waldo. We were all sitting on high stools at a lunch counter. I had put the precious bottle in my pants pocket, and as I rose up to sit on my stool, the bottle slipped out and loudly crashed to the tile floor, splattering pink liquid *everywhere*! It was sensational! But I was not only chagrined... I was desperate!

One Answer From Antiquity

Since none of us had consumed even a drop of Pepto-Bismal – and we were all very anxious to hopefully get started on recovery, I didn't take time to order any food. Immediately, I took off in a brisk walk back several blocks to the drugstore to replace this loss.

Try to imagine the drugstore clerk's reaction – as he saw me re-appear in only minutes to ask for *another* bottle of Pepto-Bismal. I've so often wished there had been a video of the "shock, awe, and puzzlement" on his face. I thought he might even faint!

It was impossible to explain why I needed a second bottle, since I could not speak Spanish. He obviously knew nothing about the two other desperate victims who also needed it. I couldn't tell whether he might be a diligent pharmacist who felt obliged to warn against overuse or maybe just had fear that it was being used for purposes other than legitimate – like re-selling it. Worst of all, he didn't understand English!

Slowly and very reluctantly, he brought out another bottle for me – that critical and essential *answer* . . .

Much of life consists of questions and answers. Answers are *sought*. They are either *replies* to questions or *solutions* to problems. Obviously that Havana answer was the latter – and quite different than many other answers we might receive throughout our life – in content, timing, rationale, and meaning.

Questions and Answers

One of the earliest and most interesting forms of human speech involves *questions* that children start asking as they grasp linguistic ability. Curiosity bubbles up into word format to demand answers. *Why, who, when, where, how . . .*

Questioning is a major form of human thought and interpersonal communication. The questioner uses questions to explore an issue, an idea or something intriguing. That process then forms and wields questioning to hopefully develop *answers*.

Questions always *precede* answers. But we all know that answers are not always possible. In fact, many unanswerable questions about *death* have already been discussed — perhaps the most interesting being "Is death the *absolute end* of life or the *boundary of a new state* for life?"

Countless books have been written – and continue to be written – attempting to answer all kinds of questions concerning death. Of course, every author who writes about what transpires following death is limited to being *speculative* since none of them have any first-hand experience with the subject.

Seeking to provide objective insight on the most important question about death – *what happens after death*, one ancient book is proposed as having the answer. Recommending this book for thoughtful examination should not be viewed as promoting sectarian or religious interest. Its *historic significance* alone justifies its serious consideration.

By any standard, no other book has ever been written that discusses death in the depth that this classic work does – the *origin* of death, the *reason* for death, the *results* of death, and the *consequences* of death are all discussed in it.

Year after year, its sales far exceed the sales of every other book in the world. Yet it is never listed on any Best-Seller list. Translated into over 500 languages, it is the most widely distributed publication in the world.

Why a *Super* Best Seller?

What is this magnum opus that has been blessed, banned, and burned for centuries? The Bible (Greek for "books"). Written over a span of 1500 years, it comprises the writing of 40 authors – all of whom consistently address the singular theme that God Who created life and humanity desires a *relationship* with all of us. And death plays a consistent, all-encompassing role throughout the book.

The Bible was the first book printed in the Western world using movable type and marked the start of the age of the printed book. It has an iconic status exceeded by no other written work. People of all ages memorize portions of it. Its influence on history and culture — including literature, music and the other arts — is incalculable. Quoted widely in many other works, it is loaded on untold millions of mobile devices, smart phones, and computers.

Book sales generally follow marketing, exposure, advertising or some other promotional effort. Yet seldom is any overt effort or peddling expended to *sell* the Bible. Many people even own *several* Bibles! For many generations, the family Bible was used as the only document in which to record births and deaths.

In the United States, placing one's left hand on a Bible while reciting the oath of office is a long-standing custom. It dates back to George Washington. He also *kissed* the Bible following his inauguration – establishing a tradition that continued with all Presidents up to and including Harry Truman. Dwight Eisenhower broke this observance by saying a prayer instead of kissing the Bible.

When I was sworn into office as a Member of the National Transportation Safety Board, I continued this tradition by placing my left hand on a new Bible presented to me as a gift for this occasion. But I didn't kiss it…

Why do you think that 33 successive American Presidents *kissed* the Bible? Often, they were of an opposing political philosophy and had been elected because they disagreed with their predecessor! Was that odd act a childish, superstitious gesture? What *significance*, if any, could be attributed to the Bible by honoring it so highly? Could it be that all these Presidents perceived in it a source of wisdom above and beyond political expertise? Were they subordinating their new authority to a *higher* one recorded in the Bible?

Over 500 Languages

Most books are written in their author's language. And there is a good rationale for that book-author connection. The stimulation, message, and intended audience for any book all originate with the author. True, some books are later translated into additional languages because of their wide appeal or significant message. But the assignation of original authorship is never lost.

A notable feature of the Bible is its *diversity of authorship* – 40 authors writing from within Hebrew, Greek, and Roman cultures but focused on one singular theme. So the normal book-author connection is less significant for the Bible than its *message*. Undoubtedly, the stimulus for translating the Bible must be far different than other translated books.

However, there is no other book that has been published in as many languages as the Bible has. That fact would seem to affirm that its unique message is near-universally recognized as significant and worth knowing.

Longest *Life* of Any Book

The *Library of Congress* —- facing the east side of the US Capitol in Washington — is America's oldest federal cultural institution. It is also the largest library in the world, with more than 160 million items on 838 miles of bookshelves.

As one enters the Great Hall of the Library of Congress, two monumental Bibles face each other. One, the Giant Bible of Mainz, signifies the end of the *handwritten* book. The other, the Gutenberg Bible (dating to 1454), marks the beginning of the *printed* book and the explosion of knowledge and creativity it would engender. This exhibition explores the significance of those two Bibles. And there are sixteen *additional* selected Bibles in the Library's collections.

Why do you suppose the Bible has been given such honor among all other books? What causes the Bible to be regarded as the most important book in the world? How and by whom was the Bible selected for this premier status?

Even more amazing than its *historical* status above any other book is the Bible's *unbroken production* to this day! It continues to live on – with nearly 600 years of publication.

Not All Owners Are *Believers*

Despite the existence of limitless numbers of Bibles in the world, it may be surprising to realize that many of the people *possessing* a copy neither read nor believe the Bible. Numerous Bibles are simply retained, honored, and viewed as beautiful library volumes. Other copies may be considered to be of cultural or historic significance. Yet the value of the Bible lies not in *owning* a copy but in comprehending, believing, and adopting its *revolutionary message*. It is a radical life changer!

Whether the basis for the unusual Presidential gesture of kissing the Bible is ever resolved, there are likely innumerable reasons for the Bible's unusually long-standing veneration and respect. However, there is only one reason for proposing the Bible as having the unique answer concerning death. It is to expose its radical, systematic, and thorough treatment of that otherwise elusive subject for which there are no other confirmable answers.

Reservoir of Death Resources

It may be surprising to recognize the primary virtue of the Bible – and the rationale for offering it as a solution for unraveling of the riddle: *what happens after death*. The Bible's entire objective is to provide a means for *guaranteeing life after death*! Though it is comprised of history, poetry, tales of warfare, proverbs, predictions, and revelations, its overarching intent is to assure that death can be conquered instead of the inverse — death conquering everyone. It is loaded with practical tips on overcoming the sting of death.

There is one central message consistently addressed by all 40 authors of the Bible: God created the human race with the express – and single — purpose of establishing an *eternal relationship* with humanity. Though the first human beings deliberately broke that relationship – thereby incurring the known and acknowledged penalty of death, God has always provided a means of restoring that broken fellowship.

The Bible is simply the recounting of that ongoing compassionate restorative process. It assures us that God is continuing to pursue humanity, encouraging us to enter an *eternal* relationship with Him — to know, trust, and communicate with Him.

The Bible not only inspires us, it explains life and God to us. It does not answer *all* the questions we might have, but enough of them. It shows us how to live with purpose and compassion – even how to relate to others. It encourages us to rely on God for strength, direction, and enjoyment of His love for us. The Bible also tells us how we can have a life and identity that – while experiencing physical death – will never end!

Broad Spectrum of *Authors*

One of the outstanding features of the Bible is its authorship. The 40 authors not only lived over a span of 1500 years. They also lived in different *areas* on earth – under diverse forms of government. Some were politically powerful. Others were persecuted – living at the bottom of society.

Even the education of authors – equipping them to write – was widely varied. Moses, for example, was obviously well-educated in the household of the leader of the most powerful nation on earth. Some authors are often unidentified so there is no way to know their educational background.

Others – like Saul of Tarsus, later known as Paul, studied under the leading educator of his time.

The terms and conditions under which the authors *died* are also widely varied. Moses died at 120 years of age, after climbing a mountain while in excellent health. Joshua likewise finished his life in good health following a successful career. Paul, in contrast, was crucified after being held a long-time prisoner in a foreign country.

Diversity of *Cultures* (Hebrew, Greek, Roman)

Another unusual aspect of the Bible is its wide cultural foundation. Culture can be defined as all the guiding values, outward signs, and symbols collectively taken together and held as important by a large group of people over a long period of time. Individual cultures develop their own customs – rituals or other traditions that express outwardly a group's cultural values. And those values may not be readily obvious. They run deep!

The commonly held standards of what is acceptable or unacceptable, important or unimportant, right or wrong, workable or unworkable in a society vary over time. So the fact that the Bible has proven relevant for many centuries by addressing the unusual breadth of cultural values that have been passed down for generations suggests that it is without competition among all other books. Death, because it is universal, is imbedded in all cultures – but not always viewed in the same manner. However, the Bible uniquely removes the mystery of death for all cultures. It is *supra-cultural*!

Only Demonstrated *Resurrection*

Certainly the most distinctive feature of the Bible is its historic proof of what happens *after* death. And there are numerous recorded eyewitness accounts of a person being physically resurrected following a brutal execution that was witnessed by a large crowd.

That proven resurrection obviously provides the required "missing link" – bridging between a living human being *prior* to death and the same living, recognizable person *after* their death. Do you recall Debi's realization about resurrection described in Chapter 3?

> "Beyond believing that God exists, I realized that it was the resurrection of Jesus Christ that we celebrate every Easter that I had to affirm in my mind – to believe really happened. After all, if He was raised from the dead by God's power, then there is life after death. That realization did it for me. I was no longer afraid to die but, in a sense, I had already passed from death into life!"

Of course, that historic fact – so widely witnessed and documented – has been a "provocative sticking point" throughout history. And for good reason. Should it be true, it raises a host of significant and agitating questions:

- Is resurrection universal for *all* human beings?
- Are resurrected beings exactly the *same people* who died – regardless of what type of life was lived on earth?
- How critical to resurrection is bodily integrity — or the lack of it; e.g., cremation — following death?
- What *changes*, if any, does a human being undergo upon resurrection?
- Is there any *accountability* for how life on earth was lived prior to death?
- How does a person *qualify* for resurrection – if qualification is required?
- Will all resurrected beings reside *together* eternally?
- Can human beings do anything, prior to death, to *prepare* for resurrection?

Setting the Stage for Resurrection

Not everyone is inclined to ponder what resurrection *means* – or why *preparing* for it might be advisable. So much of everyone's life seems to be lived on a "moment by moment" basis. After all, Darwinian "survival of the fittest" often governs day-to-day living. The dynamic ebb-and-flow of daily life is quite consuming. It demands our attention – so much so that we seldom pause to weigh or contemplate truly significant factors of our existence.

Consider two frequent aspects to this consuming mindset. First, death is continually pushed out of awareness – both consciously and subconsciously. Second, few people seem to be aware of the *possibility of resurrection* – because they have neither seen a resurrected person nor heard about resurrection. As Will Rogers said, "All men are ignorant, except in certain subjects."

The Systems Approach to Resurrection

To apply the systems approach to the idea of resurrection, it may be helpful to do so in two stages. First, we must review and weigh the progressive nature of human life – from conception to death. Second, we must examine and incorporate the possibility of resurrection prior to death.

The flow of typical major functions of life — from birth to death — was earlier described by Figure 2 in Chapter 3. Perhaps you took time to locate where you are *now* — your current existence or activity somewhere in that diagram, regardless of your age.

However, Block 20 in that diagram – "Experience Death" – sets the stage for the possibility of *resurrection*. That term is virtually unknown today. Occasionally, we call something long forgotten or thought to be ended "resurrected" if it is revived – like a dress style or type of music. But the idea of the bodies in graves in all the world's cemeteries suddenly rising to life again defies rationality for many who have never been aware of it.

The linear view of history is challenged by William Strauss and Neil Howe in their book, *The Fourth Turning*[1], in which they propose:

> Try to *unlearn* the obsessive fear of death (and the anxious quest for death avoidance) that pervades linear thinking in nearly every modern society… Without human death, memories would never die, and unbroken habits and customs would strangle civilization.

The renowned Swiss physician, psychiatrist, and author Paul Tournier has outlined the characteristics and peculiar laws that govern life as four seasons: *spring, summer, fall* and *winter* in his book *The Seasons of Life*[2]. Listed in Appendix A, they are worth reviewing.

Once we have resolved the *nature* of our life, it would seem appropriate to give serious introspection and thought to acknowledging *resurrection* as a possibility.

Resurrection is never mentioned or discussed by news media. It is not a subject taught in schools. No one talks about it in the market square. Attorneys never include it when assisting you in preparing your will. So it must be deliberately entertained from a position of great ignorance.

Resurrection's Universality

The first surprise for some readers may be that *everyone* will be resurrected whether they like it or not! It is not an *option* reserved only for those who wish to live after death. *Every* human being is destined to live *forever,* according to the Bible.

A second Biblical surprise is that there are *two interlocked options* associated with universal resurrection. These two options are far different than the option of *seat selection* following the option to fly *first class* versus *coach* on an airliner. Why? For two reasons:

- There are two radically different post-death *destinations* for every person.
- The destination option is *fixed* from birth for every person – unless an alternate is elected *prior* to physical death.

Death Involves Duality

Almost everyone, when "death" is mentioned, thinks of a *singular* event. It can occur *suddenly* – like a heart attack or in an airliner crash. Or it can occur almost *imperceptibly* at the end of a lengthy illness. In any case, it is defined *biologically*; i.e., the termination of heart or brain functioning.

Yet there is often wide recognition that some people seem to have a vital force – an inspirational characteristic, an ability to mobilize others, or a transformational vision that is independent of their physical body. It is often considered to live on – after they have died. Martin Luther King is often quoted an example of possessing this nature.

However, the Bible portrays people quite differently. All of us possess very specific differences. Every person possesses a body, soul, and spirit – each with a different consciousness.

The **TOTAL PERSON**

From *THE SPIRITUAL MAN* by Watchman Nee

SPIRIT
GOD – Consciousness

- Noblest element of man
- Innermost area of being
- Source of eternal life
- Prior to fall, total controller
- Cannot act on body – only thru soul
- Communicates with spiritual world
- Where God & man commune
- Regeneration occurs here
- Conscience, intuition & communion
- Dwelling place of the Holy Spirit
- Utterly dependent on God
- God's intent: rule the soul
- Like a mistress who commits

SOUL
SELF – Consciousness

- Binding element in man
- Contested area of being
- Outer sheath of the spirit
- Intersection of body and spirit
- Site of personality
- Will, intellect, & emotions
- Decides between natural & spiritual
- Must give consent to spiritual
- Ideals, love, choice, decision, volition
- Very life of man
- Covets what God has not conferred
- That which sins
- Steers clear of God
- All that we "naturally" are
- Like a steward who commands

BODY
WORLD - Consciousness

- Lowest element of man
- Outermost area of being
- Outer shell of the soul
- Communicates with natural world
- Nourishment, reproduction, defense
- Defiled – only destination: death
- Sins' special sphere of operation
- Earthen vessel & earthly tent
- Members of Christ when redeemed
- Province of sin's domination
- "For the Lord & the Lord for it"
- Sin's fortress, instrument, garrison
  Like a servant who obeys

Figure 1 — *The TOTAL PERSON*

Figure 1 lists some characteristics of those three components that have unique, specific, interactive functions. Only the *body* experiences death – soul and spirit continuing to exist unchanged until the body is resurrected to be integrated again with them.

In summary, human beings are *complex personalities* – far different from dogs, frogs, or hogs.

Two *Types* of Death

Death for every one of us, according to the Bible, happens not once but *twice*! And those two deaths occur for very different *reasons*. Yet those reasons are seldom exposed or discussed in modern society – even in churches or seminaries.

Our two deaths have different causes, as shown in Figure 2. Death Cause No. 1 has nothing to do with anything we have *done* – we are born with that reason for dying. It's due to our inherited human *nature* – that self-centered perspective and behavior that starts right in the cradle and never ends. We are universally guilty of "playing God" – seeing everything as subordinate to us and our values, desires, and command.

Our TWO Deaths

MORAL DEATH	PHYSICAL DEATH
CAUSE: Our sinful nature inherited from Adam	**CAUSE:** Specific sins we committed in our body
EFFECT: Eternal separation from God	**EFFECT:** Funeral followed by burial
SOLUTION: Restored original relationship to God by accepting Jesus Christ's death and substitution for our nature (spiritual rebirth)	**SOLUTION:** Resurrection with new eternal bodies
TIMING: Restoration must precede physical death	**DESTINATION:** (1) Eternal joy with God and the morally restored, or (2) Eternal separation from God

Figure 2 – *Our TWO Deaths*

Death Cause No. 2 is the penalty we must pay for the innumerable violations of right-living (called "sins") that we all readily recognize yet are unable to stop committing. For some of those violations, we may be penalized by law. But most go unaccounted, though often regretted.

Those two *types* of death do not occur simultaneously for anyone. They are separated in both *timing* and *sequence* – two very critical factors.

Figure 3 – using color coding – enables you to recognize the importance of *both* deaths. All humanity begins life in the same condition – shown proceeding from the left in orange.

Those who choose to die their moral death *prior* to their physical death – as both Debi and I have done — follow the green pathway. Those who fail to die morally prior to dying physically follow the blue pathway – their *physical* death will be followed by their *moral* death.

The early Christian theologian and philosopher Augustine of Hippo recognized these two vastly different deaths. In his book *DEATH AND AFTERLIFE: Perspectives of World Religions*[3], author Hiroshi Obayashi says:

> "Augustine differentiates two kinds of death . . . one is the separation of the soul from the body. . . while what he calls 'second death' occurs only with the Final Judgment."

Obayashi confirms Augustine's conviction that the two *deaths* are aligned or correlated with two types of *resurrection*:

> "Just as he differentiated between the first and second deaths, he now distinguishes two concepts of resurrection. By the first resurrection Augustine refers to the regeneration that is effected by means of baptism in the current life. It is the resurrection according to faith. The second is the bodily resurrection that is to take place on the Judgment Day at the end of time. Thus Augustine, by assigning the second, bodily resurrection to the Judgment Day, highlights the importance he attaches to the first resurrection, *which is to happen to us in our current life*. It is the resurrection from death (the life lived in sin) to life (the life lived in God's grace)." (emphasis added)

One Answer From Antiquity

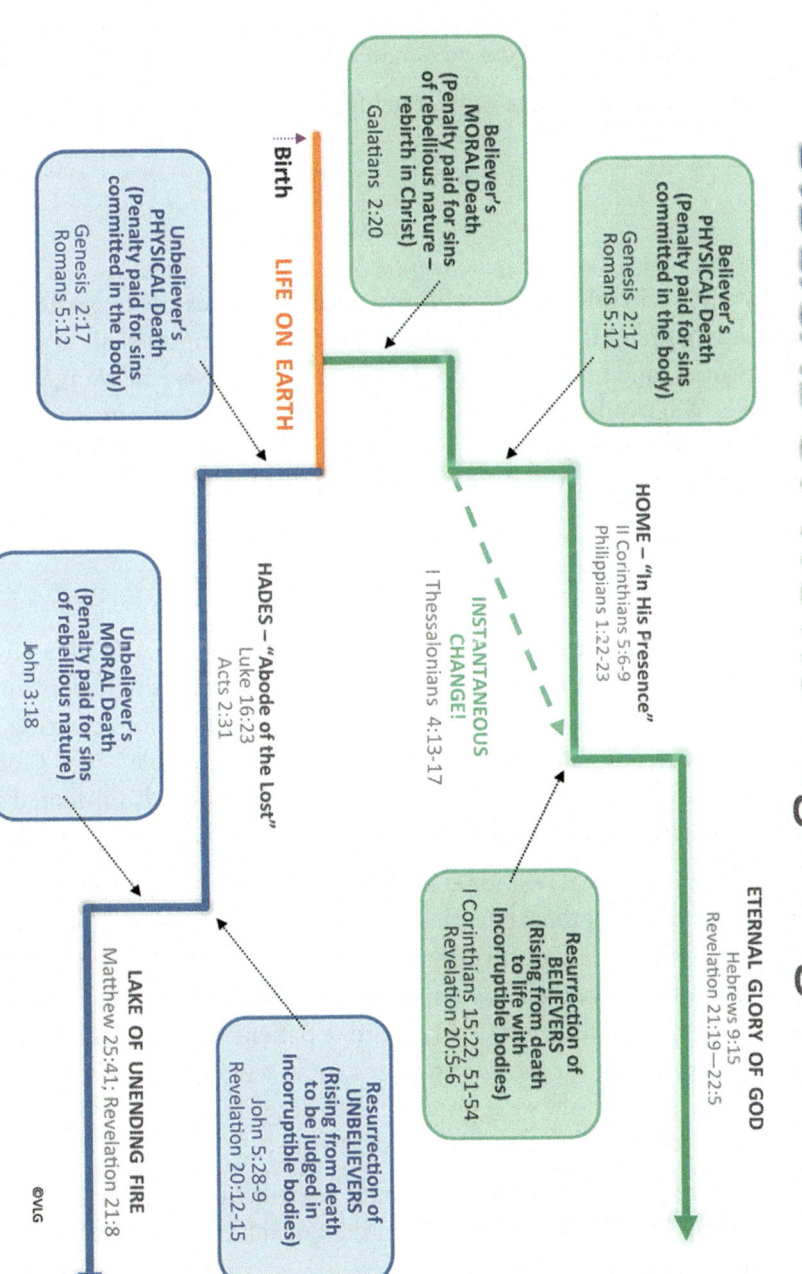

Figure 3- *BIBLICAL OPTIONS Regarding Death*

Singer-entertainer Pat Boone – a longtime friend, surprised me when he learned I was writing this book. He excitedly told me of a ballad *"Everybody Dies"* he wrote years ago based on two deaths for everyone (see Appendix B).

Some of the lyrics of that song explicitly describe those two different deaths:

Born just once, and you'll die twice.

Born twice, you just die once.

But Pat's wistful yet insightful song only broaches the incredible role that death has played throughout history for great composers of the classics. For centuries, one tumultuous theme has provoked overwhelming majesty of music. What subject is the focal point of the greatest musical masterpieces of such towering titans as Bach, Beethoven, Brahms, Mozart, Haydn, Tchaikovsky, Wagner, Verdi, and Handel? **Death**!

Two *Times* of Death

The earlier-stressed central message of the Bible is that God is vitally interested in establishing a *loving relationship* with every single person. To that end, He provides options, opportunities, and offers for restoring the broken rapport He once had with humanity. It was people – not God — who not only broke off the relationship with God but also infected the entire human race with moral death… and they knew what they were doing.

So the most critical relational bridge for all of us is overcoming the penalty for our personal *moral* death – the one we inherited without choice. Amazingly, God paid that horrible price for us – by transferring it to His Son whose death God accepted as a substitute for *our* death! However, *we must accept that gift*! The Bible says, "The Lord is patient with you, not wanting anyone to perish but for *everyone* to come to (acceptance of the substituted death for ours)."[4] Should we ignore or fail to accept this wonderful gift, the default is automatically our moral death.

In Figure 3, *acceptance* of that gift is represented by a vertical green line going up from "Life on Earth" and the green block marked "Believer's MORAL Death."

The second consequential time of death in Figure 3 is also a green block "Believer's PHYSICAL Death." That marks the time of your second death, and it pays the price for your "violations of right living" or sins.

Critical *Order* of Dying

Figure 3 clearly depicts – by color – two very different paths progressing from left to right. The green blocks represent those who elect to accept God's offer to substitute His Son's death for their own moral death. The blue blocks represent those who either overlook, ignore or refuse that incredible offer.

Augustine emphasized the criticality of resurrection timing. Given that everyone dies twice, the *sequence* of those two deaths is highly consequential! Why? Because eternal destination is determined thereby! The penalty of *moral* death must be paid before the penalty for *physical* death. If payment is reversed, the destination is fearsome indeed.

Individual Choice is *Primary*

Interestingly, the Bible discusses human death in such great detail and from so many perspectives that a systematic approach to the subject demands that its rich reasoning at least be consulted and summarized. Many religious beliefs, churches, sects, doctrines, cults, factions, and organizations are centered – even based on — the Bible. In the end, its value in answering what happens after death will depend on each individual who studies it.

This ancient answer to ageless questions is worthy of study and objective examination without any religious interpretation — simply by consulting the specific Biblical references in Figure 3.

Good News! You can *pre-plan* your own resurrection!

Chapter 9

Political Solutions for Death

"The partisan, when he is engaged in a dispute, cares nothing about the rights of the question."—Socrates

It was a bright sunny November day. With two other Northrop executives, I had awakened very early to drive 50 miles — from the San Fernando Valley, passing through Los Angeles — to attend a 9:00AM meeting in Downey, California.

Space Division of North American Aviation – the prime contractor for the Apollo Command Module – had called a meeting of 200 contractor executives to be briefed on a new concept called *Design Review*. We three Northrop representatives were responsible for creating the Apollo Earth Landing System – the three-parachute combination that deployed after reentry, then slowed and landed the Command Module carrying the 3 astronauts in the ocean for helicopter rescue.

Design Review was a technique being touted as a unique approach for integrating the technical and management complexities of America's manned spacecraft intended to hopefully win the race to put men on the Moon ahead of the Soviet Union.

The NAA Space Division Vice President greeted everyone in the large auditorium before introducing a lengthy video presentation that defined and illustrated step-by-step how design reviews were to be conducted in the Apollo program by all contractors.

At that time, putting men on the Moon was still a cross between dream and reality. Many unknowns were yet to be solved. Skepticism vied with optimism as we watched the video.

After a coffee break, we all once more took our seats to watch the second hour on Design Review...

Suddenly and unexpectedly, the video stopped – and the famous Albrecht Dürer "Praying Hands" painting filled the screen!

"Praying Hands" by Albrecht Durer

The auditorium erupted in wild laughter, whoops and guffaws! We all assumed that the projectionist had cleverly found some wacky humor to break up the intensity of the Design Review presentation — and to perhaps lighten its approach for addressing obvious Apollo challenges just ahead...

But almost immediately, the scratchy sound of patched-in TV audio blared, "Dallas, Texas...." Then audio went silent. Subconsciously, I was aware that President Kennedy was flying to Dallas from San Antonio that day. The thought popped into my mind, "Maybe the White House has just made an important announcement."

Within a split second, the recognizable voice of Walter Cronkite loudly boomed, "President Kennedy has just been shot!" Somehow – and incredibly, the projectionist had managed to connect CBS television coverage into the auditorium!

The raucous laughter *instantly ceased* – just as rapidly as it had exploded. I noticed most of the executives suddenly bowing their heads in silence – responding to that brief visual of praying hands and Cronkite's curt declaration.

In my entire life, I have never witnessed such a remarkable transformation from wild exuberance to sober concern – even to prayer! It was as though a military *command* had been given during close order drill, resulting in instant unanimous response!

The video was turned off. The Space Division VP went to the stage, suggesting that we all pray for Mrs. Kennedy. (I've often wondered why he didn't also mention prayer for the President because Cronkite had not mentioned his death.) Nonetheless, he then abruptly adjourned the conference without any summary or closing message.

As many of us shortly learned — while having lunch in a nearby restaurant, 46-year old President John Fitzgerald Kennedy was *dead*.

The visionary for our Apollo meeting that morning was gone! Only 14 months earlier at Rice University in Houston, he had made his famous declaration: "We choose to go to the Moon in this decade" – acknowledging that it was not because it was easy but because it was hard!

Politics — Expectations and Results

The assassination of a President or any head of state immediately pre-empts all other public interest and focus. We all likely recall that in 1914 a Serbian teenager assassinated Archduke Franz Ferdinand and his wife as their motorcade drove through Sarajevo – quickly setting off a chain reaction of events culminating in the outbreak of World War I.

The unexpected death of a President "sucks the oxygen out of the room."

In a complex world – and especially in the United States, political convergence is on the power of numbers, fierce competition of interests, winning elections, and managing opinion.

Immediately, there is concern about leadership – particularly how to maintain continuity in building and keeping unbroken the trust everyone had in the previous leader. It is far different than the transfer of power via inauguration following a long election campaign. The successor is generally unknown, even though elected along with the now dead President.

All Americans place some faith in the political process – especially in representative constitutional government where policy is not only *enforced* but the mechanism for determining that policy is also *determined* by the governed who choose to follow political leaders.

Periodic elections at many levels are always initiated by campaigns conducted by numerous competitive candidates for office — enumerating and promoting lots of promises to provide needed change, if *they* are elected. Billions of dollars are expended in the United States on these campaigns. Reputations are created, attacked, defended, and destroyed in this circus-like activity.

But *results* seldom match *expectations*. Those elected appear to either forget what they promised or are unable to deliver it after assuming office. Why is that true? Are politicians deliberate *liars*? Are they *naïve* – before gaining office – about their ability to carry out their promises? Does the political process somehow *poison or preclude* office holders from carrying out their promises? Bottom line: Is promise of political change an *oxymoron*?

Of course, we all seem to accept the promise-but-no-delivery character of politics. Why? Likely because everyone recognizes that the centerpiece of elective politics is *compromise* that is required for two reasons. First, the population is *diverse* not uniform – holding many views on any given subject. Second, that diversity virtually precludes *unanimity* on any issue.

To confirm the low probability of political change, consider the Constitution of the United States. There have been 11,539 proposals to amend the Constitution introduced in Congress since 1789. Yet only 33 amendments have been adopted by the United States Congress and sent to the states for ratification since the Constitution was put into operation on March 4, 1789. That's a batting average of 0.00286 – less than 3 successes out of every 1,000 attempts!

Baseball would have neither charm nor allegiance if batters had that batting average!

Danger of Political *Overreach*

Despite the widely known hype and hypocrisy of politicians, people continue to turn to them for solutions. In the United States, the political influence of K Street lobbyists in Washington, DC continues to grow – not only numerically but financially in promoting special interests. So why should it be remarkable that laws and regulations continue to proliferate out of Capitol Hill at an exponential rate?

Congress has been passing new laws for well over 200 years. One would think that legislators would eventually figure out *all that is necessary* for governance. Will writing new laws never *end*? Why *not*? Is the populace so chaotic, confused, or committed to wrong-doing that current laws will always be so sufficiently inadequate that *new* ones will have to be *continuously* written?

Of course, laws are also increasingly being written – not to *correct bad behavior* but from a philosophical position of paternalism. These laws empower *government* to usurp responsibility for decisions that historically have been made by *individuals* – and based on the proposition that government knows best. Examples include health care, contraception, toilet design, education, and financial compensation.

Although not always recognized, new laws do not stand alone but often also result in creation of a plethora of associated *regulations* intended to *interpret* those laws. Collectively, these regulations become a runaway governmental force that both the executive and judicial branches are unable to control. This authoritative power is cumulative – and never reduced.

Theoretically, law is the activity of subjecting otherwise independent human behavior to the governance of *rules*. So the rule of law is concerned with regulating the use of *power* to attain behavior considered acceptable to a majority of those being governed. But how many laws are *required*? Will legislators *ever cease* writing more laws?

While serving Ronald Reagan as a Member of his 5-man Governor's Select Committee on Law Enforcement Problems, I briefed him on the negative impact that the increasing rate of new laws had on *enforcing* all laws in general. As the number of laws increases, the need to assure *compliance* also increases. But in fact, there are thousands of laws and regulations that are not — and never will be – *enforced!*

One solution I proposed to Governor Reagan – and which he found worthy of consideration – was to urge passage of a law by the legislature that would require repealing or abolishing *two existing laws* as a condition for passing any *new* law.

Legislators would likely oppose this concept because it would force revision of the legislative process in four ways:

(a) Require research to *identify* extant but no longer meaningful laws for repeal

(b) Repeal and link two extant but obsolete laws to every proposed new law

(c) Purge and reduce the law library *perpetually* as new laws were created

(d) Force a *focus* on enforcement criteria to assure effectiveness of every new law

Though my proposal was never adopted, I believe that it still has merit!

Political life is anchored to law – defining, regulating, and enforcing behavior. That authority is intoxicating business, too. As Lord Action once said: *"Power tends to corrupt and absolute power corrupts absolutely. Great men are almost always bad men."*

Most politicians believe that they have a unique and significant role in solving all types of society's problems – including those involving death. And they are correct – when assuring national defense as well as protecting citizenry physical well-being from harm.

However, politicians are – and always will be — limited in their authority and capability. And they must be watched for exceeding proper boundaries when colliding with moral issues involving death; i.e., legalizing *murder* of the unborn and the infirm or declaring the indeterminable *motive* of mass murderers. The recent expansion of political control to assume the patronizing role of ever-increasing specification of a citizen's legitimate food, exercise, body weight and reproduction should be of constant concern. It is only one step short of attempted *thought* control.

Laws Seldom *Solve* Problems

The politician's primary means of control is the law. Postulating a need, negotiating a solution to that need, soliciting and obtaining sufficient legislative support for adoption of that solution, defending the adopted law, and taking credit for its success. That's the five-stage life cycle of successful politicians.

The most dramatic use of political power and legislation is provoked by real or perceived *apocalypses* — sudden and massive disasters or very bad situations that could cause incredible fear, loss, death or destruction.

A current example of this political phenomenon is the concern about *global warming* that was credited in Chapter 6 with launching Earth Day in 1970.

Ponder whether there is any apocalyptic potential in this common definition of global warming:

> "A *gradual* increase in the average temperature of the earth's atmosphere – *generally* attributed to a greenhouse effect caused by increased levels of carbon dioxide, chlorofluorocarbons and other pollutants that are sustained *sufficiently* to cause climatic change."

Note the *ambiguities* – uncertain, inexact, vague, indefinite, and unscientific. That's one of the hallmarks of an apocalypse for ambitious politicians. *Death* or *extinction* seems to be imminent unless laws are passed. It is fertile and ideal for marshalling political support by its alarming specter – yet with no accountability for *success*. There are neither means of proving it *exists* nor if it is ever *removed*! Weasel words win.

This particular apocalypse even attracts *theologians* like Roman Catholic Pope Francis who issued an encyclical warning of its possibility!

However, its alarming possibility is *ignored* by the world's worst polluters — those contributing the majority of carbon dioxide, chlorofluorocarbons and other pollutants. That lack of global involvement obviously precludes any meaningful mitigation of the apocalypse. So it lives on ... a politician's goldmine.

To compound the irony of this phantasm – and in spite of its unlikely global resolution, US political advocates continue to support legislation to eliminate or modify industrial processes and penalize through regulation the utilization of natural resources to limit or circumscribe life style.

Consider these amazing *facts* about the earth's temperature:

- It is never a *singular* value – varying widely all over the earth
- It varies cyclically — constantly *changing,* both lower and higher
- It correlates to mostly *external,* non-earth factors – primarily activity on the sun
- It has no proven connection to *human* activity
- It has no "normal" or *ideal* value to be sought (if it were humanly possible)
- It was unknown or estimated prior to 1880
- It has no known impact on duration of human life

Of course, there are other current US apocalypses that also attract political involvement such as universal handgun ownership, unrestricted free speech, and unlimited sexual anarchy.

Frédéric Bastiat's classic essay, "The Law" — first published in 1850 by the renowned French economist and journalist — provides a concise statement of the original American ideal of government. Based on the Declaration of Independence, Bastiat believed that all human beings possessed the God-given, natural rights of "individuality, liberty, property" and that these "three gifts from God precede all human legislation."

However, even in the 1840's, Bastiat was alarmed over how US law had already been "perverted" into an instrument of what he called *legal plunder.* Far from *protecting* individual rights, the law was increasingly being used to *deprive* one group of citizens of those rights for the benefit of another group, and especially for the benefit of the state itself.

He saw that such plunder would erase from everyone's conscience the distinction between justice and injustice. The plundered classes would eventually figure out how to enter the political game and plunder their fellow man. Thereby, legislation would never be guided by any principles of justice, but only by brute political force.

Hate Is A *Crime*?

A stark example of the irrationality of expecting laws to *solve* problems is the recent classification of law called "hate crime." First passed by 30 States and the Federal government in the late 1980's, these laws seek to solve crimes perceived to be motivated by racial, religious, gender, sexual orientation, or any other *prejudice*.

Everyone experiences hate – both self-generated and as a recipient of someone else's. Its intense hostility toward and/or aversion from people, things, foods, athletic teams, colors, or habits is common. We've all said "I just *hate* _____" and meant it. Extreme dislike or antipathy is a universal experience. So it seems rather odd that a very common human trait should suddenly become – after centuries — a special motivator of *crime*.

Traditionally, any crime has been an *action* deemed injurious to public welfare and legally prohibited (known as *actus reus*). In addition to that violation, many serious crimes require the proof of *intent* to commit it (known as **mens rea**) before a person can be convicted. In other words, the prosecution must prove not only that the accused *committed* the offence but they did it *knowing* that it was prohibited. On the other hand, many statutory or regulatory offences do not require proof of **mens rea**.

Hate crime is an attempt to expand that historic definition by seeking the *motive* for prohibited action. Doing so, it moves into untried and unproven territory – filled not only with inability to conclusively prove *any* motive "beyond a shadow of doubt" but also encroaching on the accused's right to privately hold personal *beliefs* on any subject without jeopardy, public exposure, or violation of current "political correctness."

If *hate* is a legal *motivator of crime*, are *envy, lust, jealousy, anger, bitterness,* and *passion* soon to become categories of crime motivators as well?

The necessity of *judging* the whole range of values that an accused citizen may privately hold before selecting the one deemed responsible for a specific crime is odd – even frightening — indeed. The celebrated stockbroker Bernard Madoff was charged in 2009 with securities fraud and sentenced to 150 years in prison — without any search of *his motive* based on any of his personal beliefs. Should it have called for another new form of law — *greed* crime?

When charging a person of a crime, why is *motive* only *selectively sought*? Are there *two* types of crime wherein motive is only relevant for *one*? Is this even-handed justice?

This remarkable incursion into an accused person's *thought life* – beyond their actions – readily raises serious questions about US jurisprudence in recent issues of massive death like terrorism, workplace violence, school shootings, and open place bombings.

- Who *decides* what values, thoughts, and motives are right or wrong—and on what basis?
- How can a person's *thoughts* be definitively known "beyond a shadow of doubt"?
- What is the *source* of illegal thoughts and values?
- Should the source of illegal thoughts be *pursued and eradicated* by police?
- How does First Amendment "freedom of speech" protect, if at all, *illegal* thought?
- How can a citizen know they are thinking *right* thoughts or holding *acceptable* values?
- When will *thought police* become the next means of achieving law perfection?

Given the accelerated acceptability in America of using psychedelics (drugs intended to affect the mind in ways that result in the experience being qualitatively different from those of ordinary consciousness) – coupled with and increasingly stimulated by graphic mayhem portrayed in movies and video games, *a culture of increasing violence* seems inevitable. Psychotic behavior – often linked to hate crime — frequently results from consuming many *legal* pharmaceutical drugs prescribed for relief of mental torment.

Are laws *effective* in leading society in the right direction to solving its problems? Cynics would say that laws only force people to be more creative in *violating* them!

Political Bookends on DEATH

Most politicians believe that they have a unique role in solving all types of *societal problems*. This belief — coupled with political power generated by access via electronic media to instant public communication — frequently propels government officials, executives, legislators, judges, and regulators into issues well beyond their ability and authority.

Human death is one of those issues. At both ends of the spectrum of human life, government wrestles clumsily with even *defining* — let alone *creating, executing* and *interpreting* — regulatory laws on such volatile issues as viable life, dignified death, killing v. consensual death, premature death, physician-assisted suicide, right-to-die, right-to-live, and "execution as reasonable death" mode.

In all those political decisions, death is treated as *finality* – the end of existence.

Referring once again to Figure 5 in Chapter 1, political solutions regarding death resemble *bookends* where laws influence both the known inputs and desired outputs of human death.

Laws Precluding Birth

There is an increasing proliferation of laws in the US that impact the initiation of human life – even *before* conception. Under the guise of *health* legislation, all American taxpayers are required to pay for *contraception* – preventing genesis of life while removing all responsibility for sexual intercourse and thereby opening up and approving unlimited sexual indulgence while reducing the birth rate.

Following conception, *abortion* (slaughter and dismemberment) of living fetuses is also authorized by law that ostensibly protects the mother's health and well-being. Unbelievably, this has resulted in marketing body parts of aborted unborn children – even using partial-birth abortions to supply intact body parts. Even more barbaric is selecting and utilizing – prior to murder — specific techniques for crushing the infant's skull to obtain the maximum number of desired organs for sale.

A third legal avenue for precluding life is the 2015 Supreme Court declaration that *marriage* in the US – the historic source of human life — no longer consists of two humans capable of producing life. Instead, marriage is viewed as a self-oriented contract to obtain preferred tax benefits and no longer a fundamental unit of society.

Laws Avoiding Death

Once life begins, lawmakers begin. They produce almost unimaginable types of laws to govern behavior, health, education, food, travel, housing, plumbing, communication, and... *death*.

Many of these statutes have been intended to warn against or prevent citizen death. The foundational law, of course, is the US Constitution that declares: "The Congress shall have Power to... provide for the common Defence and general Welfare..."

From that broad base, *national defense* against death is provided. And public *welfare* has been broadly interpreted to include laws governing *safety* (transportation, vehicles, buildings, energy, and food), *health* (medicines, drugs, vaccination, hospitalization, and diet), *employment* (pay scale, work conditions, and bargaining), and even *clothing* (labeling, quality, material, and dyes). All of these laws could be considered as either intended to delay or prevent premature death by maintaining health.

Unintended Consequences

There are always *consequences* to laws – beyond the inherent penalties for violation and associated with enforcement. They create desired *compliant* behavior as well as deliberate intent to *violate* them – even from the moment of initiation.

Beyond those contrasting outcomes, however, laws can produce consequences that are not only unforeseen and unintended but often undesirable and counterproductive to society. Among this type of consequence in the US is a declining birthrate, decimated family structure (fatherless children raised by single mothers), increased inner-city crimes of violence, broad use of narcotics – even by children, and breakdown of law and order.

Political *Impotence*

The fundamental proof of impotence in ruling over these life-and-death profundities does not lie in the highly polarized forces created by political pontifications. Instead, lawmakers seem to be facing societal division beyond their ability to be effective. The "rule of law" is becoming a hollow concept. Such division is common on many topics related to death.

For example, there are those who favor the death penalty while opposing partial-birth abortion. Others consider death-row execution inhumane (regardless of mechanism) while readily supporting abortion as a contraceptive. Pacifists will sometimes call for increased police force against crime.

Washington Post journalist Charles Lane recently remarked on TV, "The mystery of human evil defies effective law control." Yet *evil* is an unmentionable term today in the news. Have you ever wondered *why*?

Swirling and Ambiguous *Coalitions*

Vibrant democracy has always provoked contrasting opinions and allegiances. As the campaigns supporting upcoming political elections become organized, the intensity and focus of competing interests increases. Coalitions – temporary alliances of political interests formed to support a common goal – emerge to solicit wide financial support for either toppling or raising political dynasties represented by a particular cause.

Often life-and-death issues form the heart of these controversies – capital punishment, abortion, war, euthanasia, and justifiable homicide. Opposing sides swarm like bees, accusations fly, and incredible rioting wreak widespread havoc. For all the hype generated, meaningful and lasting political resolution seldom emerges.

Unfortunately, these conflicts never lead to a thoughtful pause for pondering the major forces and substantive issues underlying them. So they remain to be resurrected when provoked by occurrence of another isolated incident.

Polls – Poor Tests of Reality

The Literary Digest is best-remembered today for the circumstances leading to its demise. As it had done in 1920, 1924, 1928 and 1932, it conducted a 1936 poll that predicted that Kansas Governor Alfred Landon was likely to be the overwhelming winner. *It had always correctly predicted the winner.*

That 1936 poll showed that the Republican candidate would be a runaway winner over President Franklin D. Roosevelt. However, in November, Landon carried only Vermont and Maine while President Roosevelt carried all the other 46 states. The magazine was so discredited by this discrepancy that it soon folded.

Even though *The Literary Digest* had polled 10 million individuals (of whom about 2.4 million responded — an astronomical total for *any* opinion poll), its polling technique was obviously faulty. And that is always a possibility for pollsters. There is no perfect way to determine public opinion. Politicians have no accurate means of testing the validity of their views – aside from election.

Covert Belief Reigns – *Regardless*

Political solutions for issues impacting death will always be at the mercy of the citizenry in a democratic nation. Since laws must be obeyed to be effective, obedience is required. But what causes anyone to *obey*? Good manners, fear of police, strong conscience, concurrence with a law's intent... there may be other reasons. However, at the root of all law effectiveness is the belief system of each citizen.

Belief is not a religious idea. *Everyone* believes. Most folks believe that the earth will continue rotating, water flows downhill, and everyone will ultimately die. Yet there is wide diversity of belief in many other subjects. Blaise Pascal – the French mathematician, physicist, and philosopher – once said, "In faith there is enough light for those who want to believe and enough shadows to blind those who don't."

In the recent past in the US, the concept of "political correctness" has emerged as a tacit but expected behavioral standard. This hyper-sensitive cultural suppression is very inexact, imprecise, and unavailable in published handbooks. It is not taught in school. It is sporadically enforced for convenience by those offended by what they personally and arbitrarily consider a violation committed by an opponent.

Significantly, it is the *antithesis* of "free speech" protected by the First Amendment to the US Constitution.

Politicians must remain aware that there is a limit to "things that are amenable to law control." Mental illness – however defined — will remain a constant problem for law enforcement. Motives are elusive, private, and essentially indeterminable.

Most of all, when dealing with life and death, what citizens *believe* will always outweigh the authority of political solutions.

Chapter 10

Toward a Rational Forum

"A thing that nobody believes cannot be proved too often."
– George Bernard Shaw

"You'd better get back down on the street *soon* – the media has massed and is demanding that you give them an interview for their critical 6 o'clock news deadline!"

Les Reingold – National Transportation Safety Board "Go-Team" press agent — was anxiously insisting that I abandon the tangled site of a head-on collision of two Amtrak passenger trains in New York City simply to meet the press.

His persistence may sound appropriate, on one hand, because I was the NTSB Board Member heading the investigation and the only spokesman. But on the other hand, you might appreciate my reticence.

This ***fourth** major Amtrak crash during July 1984* had occurred on July 23rd at 10:45AM. I had mobilized the "Go-Team" shortly after being notified at 11:50AM in Washington. Les and I – along with other team members – were aboard a flight to New York by 3:00PM.

Two Amtrak passenger trains had *collided head-to-head* in broad daylight on a viaduct high above a street in the Queens borough of New York City. One person was killed and 125 were injured – 8 seriously.

Incredibly, Train 151, pulling five coaches headed *from* Boston to Penn Station in Manhattan smashed into Train 168 pulling seven coaches — headed *to* Boston — that had just left Penn Station on its way from

Washington. The site of the crash was a steel-and-stone viaduct running from the Sunnyside rail yards in Long Island City section of Queens, through Astoria to the Hell Gate Bridge. Both engines and four cars on each train were derailed.

The wrecked trains were barely visible from the street – high on the viaduct. From the roofs of nearby buildings, the red-and-white-striped, silver cars could be seen as twisted and tipped at crazy angles. In the forward cars, most of the window glass was missing and the ends of the cars, which had slammed into each other, were bent and torn.

Rescuers had difficulty reaching the elevated tracks. The most seriously injured passengers were brought to the ground on stretchers laid across the buckets of the Fire Department's 100- foot hydraulic towers, commonly known as "cherry pickers."

The only access our Go-Team had to that remote location was by being lifted over 80 feet high in the air onto the trestle from the Queens street on the same Fire Department "hook-and-ladder" (a platform or bucket attached onto a mechanically bending arm or "snorkel" installed onto a fire truck).

The "hook-and-ladder" that lifted us up to (and down from) the New York City trestle site of the Amtrak head-on crash on 23 July 1984

It was a "first" for all of us.

Taking charge of the investigation of any accident is always a challenge. Go-Team specialists in rail traffic, locomotive performance, track signaling, human factors, and train communications had to be organized and assigned analytical tasks. Examination and study of the crushed front cabs of the two electric locomotives had to be initiated. Our team had only been on the accident scene for about an hour – surveying and establishing our investigative effort.

I had begun meeting with Amtrak officials as well as Federal Railroad Administration personnel who were coming on-site. Rescue and transport of injured survivors was still underway. The accident site was swarming with firemen and police.

Of course, news reporters were unable to get access to this scene of mayhem.

So when Les Reingold – carrying out his news coverage responsibility – kept pressuring me to leave that chaotic accident scene simply to provide news reporters answers to their questions, perhaps you can understand my conflict of priorities.

And it was compounded, of course, by the fact that I had to ride that fire truck "hook-and-ladder" back down to street level to meet them.

Unless you are a firefighter, you may need a better description of this unique device for hoisting and lowering a person. These aerial ladders are integrated to an expandable hydraulic boom. They are complex. They are equipped with a control unit, lighting equipment, a fixed water way, power outlets and compressed air outlets. An additional arm gives a horizontal platform the ability to go "up-and-over" or bend over a roof.

However, I soon yielded to Reingold's persistence — stepping off the trestle onto that shaky platform to view a swarming, competitive, boisterous circle of news correspondents 80 feet below on the street. There appeared to be about 40-50 of them. They began shouting questions at me almost immediately – even as the hydraulic ladder descended! Many seemed to be waving microphones on extended sound booms, trying to get answers before I ever set foot on the ground. Frankly, I'd not previously seen such rude, inconsiderate demands for shouted questions.

I felt as though I was being lowered into a lion's den, about to be devoured. Simultaneous but conflicting questions being barked at me made it virtually impossible to provide rational responses.

Understandably, there was good reason for rapid and accurate news media coverage. Amtrak rail service to and from New England had been blocked well into the afternoon, with thousands of travelers delayed in New York and New Haven. Delays continued all day, and railroad officials warned that half-hour delays could be expected well into the next day.

I not only comprehended the need for the public to be rapidly informed but also desired to provide accurate details for public dissemination. Further, I knew it was going to take time for news media to digest and organize answers about the accident – particularly about death and injuries as well as restoration of East Coast rail service, for its imminent evening telecasts. So I tried to respectively respond to their diverse inquires for about 30 minutes before returning to the accident scene.

Later at 11 PM, I gave a 30-minute *CBS Newswatch* TV interview at the crash site. Early the next morning, Joan Lunden interviewed me on ABC's "*Good Morning, America*." I was rushed immediately afterward to NBC where Bryant Gumbel interviewed me on "*The Today Show*." By addressing all three major networks, I had met my objective of providing accurate and meaningful information concerning the Amtrak tragedy.

News Media and Death

Recounting this unique series of encounters with news media not only evoked – but continues to stimulate even today – thoughts about the major role that news media provide in reporting about death. More than any other subject, *death* – its cause, location, number of fatalities, impact, source, type, pathos and sorrow — commands top billing in the media.

Consider these aspects of news coverage in America:

- The fanatic feeding frenzy of news *gathering*
- The value of a news "scoop" – being *first* to disclose
- Use of terms like *sensational, breaking news, bulletin*
- Role of *viewer ratings* that determine TV promotional advertising rates
- Interruption of programming to insert "*urgent*," "*flash*," or "*alert*"

Underlying this pattern of interruptive news is an assumption that it is *more important* than anything that it interrupts . . . that there is *universal interest* in receiving it . . . that there is *intrinsic value and authority* in providing it.

Death tops all other reasons for this disruptive irregularity. Death news may arise from crime, war, natural disaster, accidents, terrorism, or fame of the deceased. Recall the impact of news coverage of Princess Diana's tragic crash, the collapse of the two World Trade Center towers, or the Space Shuttle *Challenger* explosion. Death stopped *everything* . . . time stood still while everyone pondered and mourned those who died.

Why does death excite such unusual attention by news media? After all, everyone will ultimately die. Could it be . . .

- To help people figure out how they can avoid death?
- To assess *blame* for death?
- To prepare lawyers to initiate *litigation*?
- To satisfy *curiosity*?
- To answer *why* death occurred?
- To warn about *means* of death?

The incredible *priority* of reporting death is astounding when you stop to ponder it. News breaks can interrupt *any* other programming at *any* time. Yet, death is really not new – in terms of being novel, unique, revolutionary or original! It happens all the time – one of the things common to every one of us. "News" – in contrast to "Olds" — should mean something unique, unusual or rare – at least, an event that happens so unusually seldom that it doesn't happen to everyone, or shouldn't.

Quite to the contrary, *death news* is recounting something that has been occurring since the beginning of history. All people die. True, each account may have a unique setting, location, cause, and environment. But what makes it "news" is *death* itself.

Power of *Irrational Response* to Death

As recounted in Chapter 3, *response* to death has become codified in the Western world. Its proper color is black. Music changes from major key to minor, from fast to slow. Auto traffic slows – even pauses. Military cadence goes to half-step and slow-salute. Aircraft fly over a burial using a "missing man" formation. Funeral parades feature a rider-less horse. Conversation quiets to whispers. Caskets are passed in honor and solemn silence. Quietness reigns in respect.

The only explanation for this inexplicable and extreme attention is apparent public agreement that death is a powerful force worthy of interrupting normal life. However, that force can even override common sense to the detriment of needed wisdom.

Consider one example. The explosion of TWA Flight 800 occurred 12 minutes after departing JFK airport in New York on 17 July 1996 bound for Paris. All 230 people on board were instantly killed in the third-deadliest aviation accident in U.S. territory. It initiated the most extensive, complex, and costly air disaster investigation in U.S. history – requiring four years to conclude a probable cause that is still widely contested.

My involvement in 170 news interviews concerning that single disaster enabled me to observe irrationality dominated by the tradition of viewing and burying the deceased.

The immediate search-and-rescue centered on the *priority* of two aspects of death. The first aspect dealt with families desiring to have one final visual contact with the deceased. The second aspect dealt with determining the cause of the explosion in order to prevent death from a future similar aircraft disaster.

Those two priorities were in conflict – primarily because the explosion occurred over deep water where aircraft wreckage and human bodies were intermingled. To solve the mystery of the explosion required an accurate mapping and determination of *wreckage location and condition* as Priority One. To rapidly *recover bodies* as Priority One meant wreckage would have to be moved – and even destroyed — before mapping could occur.

Abandon Successful Methodology to Assuage Grief

Those two conflicting priorities in TWA800 – determining cause of the explosion v. immediate recovery of victim bodies — involved more than sequence or timing. Intensive news media coverage generated *open conflict* with NTSB investigators, as understandably grief-stricken victim families demanded that bodies be recovered as soon as possible for burial.

Relatives of TWA 800 passengers and crew, as well as the media, gathered at the Ramada Plaza Hotel at JFK airport. Many waited for many days until the remains of their family members had been recovered, identified, and released to them. This hotel became known as the "Heartbreak Hotel" for its role in handling families of victims.

Victim families managed to obtain a political platform – via news media — that totally disrupted the NTSB's investigation of the accident. Vituperative threats were shouted daily — against Navy frogmen who were exposing themselves to high risk in deep water day and night. The overworked local coroner was nearly fired by politicians because — even though his primary objective was to *identify* all recovered remains rather than performing a detailed forensic autopsy — the thoroughness of body examinations became highly variable and thereby criticized.

The priority of recovering victim bodies exceeded all other priorities — even though their recovery often meant moving wreckage and destroying evidence to locate them.

Grief then turned to *anger* at delay in confirming the passenger list, conflicting information from agencies and officials, and mistrust of the recovery operation's priorities. NTSB vice chairman Robert Francis publicly declared that all bodies were being retrieved as soon as they were spotted — and that wreckage was being recovered only if divers believed that victims were hidden underneath. Yet many families remained suspicious that investigators were not being truthful or were withholding information.

When I was asked during one of my TWA 800 news interviews to explain why this odd reaction had erupted for the first time, I proposed that public opinion about the *meaning of human death* may have changed radically in recent years.

Reordered Priorities for Preventing Death

Historically, the primary objective of accident investigation is to conceive and implement action(s) intended to prevent a similar accident from ever happening again. Recovering human bodies from an aircraft crash seldom, if ever, contributes to that objective.

That conclusion is not intended to be insensitive, uncaring or unsympathetic regarding devastation experienced by families and friends of crash fatalities. Yet it is true. But significantly, TWA 800 body recovery was allowed to interrupt and defeat vital effort *to prevent future deaths* from a similar horrible crash.

In other words, grieving people unknowingly put their grief ahead of preventing another explosion that would repeat what had *caused* their grief! How ironic!

What provoked this inverted priority? The immediate focus on the *finality of death*. The ultimate *cause* of TWA800 deaths remains an unresolved mystery to this day.

Revised *Meaning* of Human Death

The American public seems to be in the midst today of revising the role that death occupies in values, priorities, and perspective. This shift may be partly due to the products of technology – world-wide electronic communication, computers, hand-held mobile devices, on-line shopping and "the cloud." Other contributors include medical science and pharmaceuticals that continue to extend lifespans. In that milieu, death has increasingly come to mean *defeat* – losing in a promising battle against dying.

No one would rationally believe yet that death can be *avoided*, but the promise is dangled tantalizingly by potential discoveries based on, for example, stem cell research. Major bioethical issues cloud that promise, of course. At the core of controversy is the definition of *life* – and when it begins.

Of course, stem cells must be alive to be useful. So the only source for them is a living human being who must die to release those cells. That "transfer of life" raises the subject of whether it occurs *naturally* (without inducement of any type – including deliberate murder or selective butchering) during the birthing process.

This topic has created in the US major bioethical issues as well as vast and profitable marketing of resulting body parts – disturbingly reminiscent of the Nazi "angel of death" Josef Mengele's experiments to benefit German armed forces during World War II. Mengele's same disregard for life and its meaning – brutal mutilation of living people when removing bones, muscles and nerves from one subject for transplantation to another without use of anesthesia – cannot escape comparison with acquisition and sale of living human organs for medical experiments.

In the near future, as the stem cell field progresses closer to the clinic, additional ethical issues are likely to arise concerning the clinical translation of basic stem cell knowledge into reasonably safe, effective, and accessible patient therapies.

The ultimate goal of medical use of body parts is the derivation and use of human embryonic stem (heS) cells that may have the capacity to *differentiate* into all types of human tissue needed for therapeutic purposes *to avoid death* – or at least postpone it significantly.

Judeo-Christian Replacement by Secular Humanism

A very important change in American society is underway. It is subtle and not openly recognized or advertised. Yet it bears incredible influence on every aspect of national life because what people *believe* determines what *happens* to them. Crime rates, family structure, standard-of-living and political leadership are products of what society *believes* – individually and collectively.

What people *believe about death* is undoubtedly based on their *Weltanschauung* — a comprehensive conception or image of the universe and of humanity's relation to it.

Chapter 4 describes the historic Western mindset – particularly in the law – of a universe in which God is related to humanity as its intimate Creator and Sustainer. This belief is commonly called Judeo-Christian and based on the existence of God as well as values, responsibilities, and accountability associated with His existence.

The founding of America was unquestionably Judeo-Christian as evidenced by these excerpts from its Declaration of Independence – "to assume among the powers of the earth, the separate and equal station to which the Laws of Nature and of *Nature's God* entitle them . . . they are endowed by their *Creator* with certain unalienable Rights, that among them are Life, Liberty and the pursuit of Happiness . . . a firm reliance on the protection of *divine Providence*."

Secular Humanism is a comprehensive body of principles for orienting a complete human life. It emancipates the individual from all traditional controls – family, church and state – thereby empowering each person to set the terms of its own life. It holds that nature (the entirety of physical experience) is all there is – asserting that God does not exist. *Nothingness* precedes and follows human life – like bookends.

Sherwin B. Nuland, in his book *How We Die* [1] quoted earlier, describes this destiny poignantly as: "The inexorable journey by which each of us is returned to the same state of physical, and perhaps spiritual, nonexistence from which we emerged at conception."

There are likely many and diverse reasons for the recent and rapid replacement in America of its Judeo-Christian *Weltanschauung* with Secular Humanism. Nonetheless, it has affected many aspects of American life and values – even America's reputation in the world.

The impact of this radical shift no longer seems shocking – even of children killing children, as Mike Huckabee described in his book *Kids Who Kill* [2]. Societal values formerly held immutable or enduring are not only challenged – they are abolished without debate.

Renowned author and thinker Os Guinness, in his book *The Dust of Death* [3], connotes it uniquely:

> "This, then, is 'the striptease of humanism' and the doorway to postmodernism, a gathering crisis of optimism, an escape from reason, a surfacing of subterranean pessimism... First, there is the strong element of surprise... The situation is pregnant with irony. There is a crisis of disbelief as well as a crisis of belief."

The most significant result of this radical but subtle divergence of worldviews can be summarized in one word: *death*. Judeo-Christian addresses it while Secular Humanism does not.

Increased Importance of *Body* Recovery

Recall the unusual public reaction to death in the TWA 800 explosion described earlier? It demanded the very hazardous recovery of 230 bodies from deep water – even to the exclusion of ever determining the *cause* of those deaths. Physical evidence required to ascertain *why* TWA 800 exploded (needed to prevent a similar disaster) was hastily destroyed to accommodate the desire for finality — physical proof that death had occurred, thereby being able to pay proper respect for the person.

That inordinate focus is a relatively recent development in US society and illustrates the influence of secular humanism overcoming Judeo-Christian values. How? First, it emphasizes the *absolute importance* of the physical human being – the unique and only entity of interest for secular humanists. Second, it ignores any rationale or reason for the *existence* of humanity. Third, it elevates the status of humanity to a God-like status with self-interest dominating everything imaginable. Fourth, it denies any human accountability for how life is lived, since life is a Darwinian survival-of-the-fittest experience.

The 2015 Germanwings 9525 suicide crash in the Alps was a stark contrast to the TWA 800 crash. First, it occurred over land rather than sea. Second, the aircraft impacted earth as an entity rather than in disintegrating fragments falling from 15,000 feet. Third, its 150 bodies were virtually pulverized and unrecognizable.

Yet an identical frantic effort was launched to recover even tiny segments that later might be identifiable as human through DNA testing. Searchers contended with high winds as well as treacherous terrain. They climbed a rugged mountain, appearing to cling to its side by their fingernails. Bit by bit, bag by bag, like high wire daredevils, they hoisted body bags of fragments hundreds of feet up to waiting helicopters. Rescuers pinned red flags on the earth whenever they discovered new fragments.

No bodies were found whole. Authorities even deployed 45 Alpine police officers to help forensics teams — not accustomed to working in mountain ravines — recover body pieces safely.

Why all this herculean effort? Likely to accomplish what has recently become popular to describe as "closure" – a feeling that emotional or traumatic experiences like loss of contact with someone precious but gone forever has been resolved. Obviously, closure is contrasted for Judeo-Christians by "mortality being swallowed up by life" (II Corinthians 5:4).

Death: Final Extinction

The increasing prevalence of secular humanism is most readily demonstrated by how the end of life is perceived. Because death is neither mentioned, acknowledged, or discussed in secular humanism, a person simply checks out or departs into nothingness at a point in time much like a dog, hog, or frog.

The secular humanist becomes a historic figure whose existence ended when their heart stopped beating. A definite finite boundary point in time marking *cessation of being* requires the undeniable confirmation that body recovery fulfills. Final extinction is thereby complete.

Transhumanism and Singularity

Another dimension that might be explored lies beyond the two competing Weltanschauungs. Various schools of thought known as transhumanism, artificial intelligence or technological singularity propose the possibility of *post-human existence*. Death would no longer exist. It would simply be considered to be a problem yet to be solved by technology.

Time magazine featured that idea as a cover story in its September 30, 2013 issue shown in Figure 1.

Figure 1 — *September 30, 2013 cover of TIME Magazine -*

Signs of Rational Discourse

The stark contrast between Judeo-Christian and Secular Humanist views of life – and particularly regarding death — is sufficient to suggest that seeking to find any means for rational discourse about its meaning, reality, significance, or consequence could appear fruitless.

Is such discourse an admirable, reasonable, or desirable goal to pursue? Are those two conflicting *Weltanschauungs* – both extant and competing in American life today – worthy of effort to understand each other concerning the primary but critical issue separating them – *death*? Or is death destined to be an irrelevancy – poised between the two?

Given the theme and focus of this book on death, a strategy to remove it as an ignored but significant factor in our culture is proposed. Several questions immediately emerge:

- What would a rational *forum* look like?
- Are we headed toward or away from rationality?
- How is rationality about *death* defined?
- What needs to "turn around" toward rationality?
- What is required to make death a *rational* subject?

In one sense, death – whenever, wherever, and however it occurs – often appears unreasonable, senseless, absurd, unjustifiable, frightening, and illogical. Even though death is universal – everyone dies, *why* should death occur?

Many reasons for death are offered. Some are philosophic. Some are medical. Others are scientific; e.g., based on the Second Law of Thermodynamics. Still others are religious. Most reasons describe the *mechanics* of death rather than the rationale underlying its occurrence.

Death means the end whereas *creation*, in contrast, describes the beginning of anything. In between those two bookends is *life*. Rational discourse between proponents of the two popular competing worldviews seems to require an answer about whether there is *post-death existence* of that which once had life.

Understandably, there will be difficulties in seeking that answer. Yet there are numerous ideas, propositions, concepts, and beliefs about life after death in addition to the Judeo-Christian doctrine described by Figure 3 in Chapter 8.

Consider a few more approaches for assuring life after death...

Cryonics is the low-temperature preservation of humans who cannot be sustained by contemporary medicine, with the hope that healing and resuscitation may be possible in the future. Cryonics procedures ideally begin within minutes of cardiac arrest, using cryo-protectants to prevent ice formation during cryo-preservation. This idea of cryonics also includes preservation of people long after death because of the possibility that brain encoding memory structure and personality may still persist or be inferable in the future. Whether sufficient brain information still exists for cryonics to successfully preserve may be intrinsically unprovable by present knowledge.

Russian Cosmist Nikolai Fyodorovich Fyodorov advocated resurrection of the dead using scientific methods. Fedorov planned specific actions for scientific research of the possibility of restoring life and making it infinite. His first project was connected with collecting and synthesizing decayed remains of dead based on "knowledge and control over all atoms and molecules of the world." The second method described by Fedorov is genetic-hereditary. The revival could be done successively in the ancestral line: sons and daughters restore their fathers and mothers; they in turn restore their parents and so on. This means restoring the ancestors using the hereditary information that they passed on to their children. Using this genetic method it is only possible to create a genetic twin of the dead person. It is necessary to give back the revived person his old mind, his personality. Fedorov speculated about the idea of "radial images" that may contain the personalities of the people and survive after death. Nevertheless, Fedorov noted that even if a soul is destroyed after death, Man will learn to restore it whole by mastering the forces of decay and fragmentation.

Physicist Frank J. Tipler, an expert on the general theory of relativity, outlines his Omega Point Theory in his 1994 book *The Physics of Immortality*. It outlines how a resurrection of the dead could take place at the end of the cosmos. He proposes that humans will evolve into robots which will turn the entire cosmos into a supercomputer which will, shortly before the big crunch, perform the resurrection within its cyberspace, reconstructing formerly dead humans (from information captured by the supercomputer from the past light cone of the cosmos) as avatars within its metaverse.

Italian physicist and computer scientist Giulio Prisco promotes the idea of quantum archaeology which reconstructs the life, thoughts, memories and feelings of any person in the past, up to any desired level of detail. Thereby the original person is resurrected via "copying to the future."

Roboticist Hans Moravec showed in his book *Mind Children* how a future supercomputer would be able to resurrect long-dead minds from the information that still survived. This information can be memories, filmstrips, medical records, even DNA.

American scientist, inventor and futurist Ray Kurzweil believes that when the technological singularity (a hypothetical event associated with artificial general intelligence wherein a computer, computer network, or robot would be capable theoretically of recursive self-improvement or redesigning itself) happens, it will be possible to resurrect the dead by digital recreation.

These diverse approaches all share the *hope* that death will be conquered by science. So the extension to unlimited life and prevention of earthly death are goals worth pursuing. They reject that there is life after death – Kurzweil describing it as *the ultimate tragedy*. All of them want to avoid what they believe is oblivion by extending current life.

None of them indicate *why* oblivion would otherwise result or whether resurrection would be available *universally* or *individually*. Most significantly, they all omit any indication of *accountability* for how life is lived, since their view essentially is a rebellion against the concept of any type of god.

The primary reason to acknowledge these radical concepts in this setting is to support the desirability and possibility of attaining open and rational discourse to bridge between the two major opposing worldviews about death.

Stripping Away the Mask of Death

Perhaps the most significant single subject separating Judeo-Christian and Secular Humanism worldviews is death. It should not be ignored, of course, because devotees of both views *experience* death. So what might be the first step toward establishing rational discourse between these competitors? Bring death out into an open forum for examination and discussion.

Such an arena for discourse on death would allow – even encourage – reexamination of strongly-held positions without fear or embarrassment. Even though the only documented proof of life after death is the resurrection of Jesus Christ (and thus seeming to favor the Judeo-Christian position), there would be no means for forcing unanimity on the subject.

In both the TWA 800 and GermanWings 9525 disasters, the obsession to recover bodies was so intense that wisdom suggests exploring the driving force behind that intensity. To what extent were those incredibly expensive efforts driven by an unrecognized but foundational worldview about death?

After all, everyone who views a deceased person's body *believes* something about that person. Some believe that it is the last time they will ever see that person. Some believe that the body does or does not resemble the person they once knew. Still others believe that body will be transformed by resurrection sometime in the future.

Therefore, open discussion of death could even reverse the current trend of "masking death" discussed in Chapter 6. Beyond the *fact* of death, the *meaning* of death deserves examination. Its mask should be stripped away!

Attaining *Community* of Shared Values

The goal of rational discourse concerning death presumes participation in exchange of thought by a reasonably substantial segment of society. To be realized, this desirable goal would also require and involve such factors as extensive time, respected spokesmen, formal debates, a variety of forums, and widely-distributed documented positions.

Beyond those challenging requirements, open public discussion about death would be difficult to achieve outside a community that shares many cultural, social, and life values. Only communities willing to subordinate their individual interests and values in behalf of others would be convinced to engage in the desired discourse on any subject for which they hold strong opinions. Yet the charm and benefit of realizing thoughtful reasoning about death offers promise of greater understanding among all parties.

Threat-Free *Consensus*

The historic tension and competition between Judeo-Christian and Secular Humanism worldviews is understandable, since they are opposed regarding many subjects, values, and objectives. Yet attempting rational discourse on death should not be dismissed or avoided without serious consideration. Death's *universality* alone points to its high candidacy for seeking consensus.

The "why" that excites news media to interrupt and override everything and everyone ostensibly allows the living to avoid the *death du jour*. "If they had ONLY... *it* wouldn't have happened."

The consensus sought by open discussion is not "group-think" or herd instinct about death. Rather, it is a thoughtful and ongoing openness to weigh the possibility of *life after death* and the ramifications, if any, associated with that possibility. No one is coerced, obliged or intimidated.

Importantly, *tolerance* must be the foundation of rational discourse. Bigotry, defined as "intolerance toward those who hold different opinions from oneself," is ruled out because it easily degenerates into an unwillingness to tolerate or accept a person who holds beliefs contrary to yours. You do not have to *like* the person with whom you disagree, but you do have to *respect and tolerate* his right to speak.

It is time to bring the subject of death out into the public domain for open, thorough, friendly examination.

Imagine *attaining consensus* without resorting to threats, litigation, or legislation!

Chapter 11

Death in the 21st Century

> *Men often applaud an imitation, and hiss the real thing."*
> — *Aesop*

"We have a real dilemma here."

Chuck O'Hollaren, Executive Director of the 1966 National Aviation Show in Los Angeles was surveying the list of those to be honored in the Show's opening ceremony. Distinguished heroes who had made great contributions to aviation were to receive handsome plaques.

The list included Ed Heinemann, noted military aircraft designer and Jacqueline Cochran, a pioneer in the field of American aviation — considered to be one of the most gifted racing pilots of her generation. Listed also were two renowned World War II military heroes who were to be the final recipients.

When assembling the list of those to be honored, the Board of Directors had overlooked the *order* in which I, as a Director, was to present the plaques — until almost the moment the ceremony was to begin. Obviously, the sequence was not going to be as complicated or sophisticated as Hollywood's Oscar ceremony leading up to *Academy Award for Best Picture*.

However, it was my questioning who should receive the *final* plaque – the most honored one — that was causing O'Hollaren's sudden exasperation. Of course, there was no ranking problem among the civilian recipients. But the two military honorees raised real potential for embarrassment. How? By forcing us to decide which of them was the greatest – the *most honorable*!

There was no question but that Jimmy Doolittle had made a greater contribution to aviation history than Curtis E. Lemay. However, during World War II, Jimmy had only attained the rank of Lieutenant General (with 3 stars) whereas LeMay went on later to hold the 4-star rank of General for 13 years – longer than any other person in US history. And in the military world, *stars count*!

Doolittle's long list of aviation "firsts" was not matched by any living person. The list included:

- First to take off, fly and land an airplane blind — using instruments alone
- First cross-country flight in 1922
- First to perform an outside loop
- First to win all the major air racing trophies: the Schneider, Mackay, Bendix, Harmon and Thompson

Most importantly from a military perspective, Doolittle had received America's highest honor — the Congressional Medal of Honor for leading the "Tokyo Raid," the first attack on Japan after Pearl Harbor.

All the members of the US armed forces – and even the President – are expected to render salutes to recipients of the Medal of Honor whether or not they are in uniform, as a matter of respect and courtesy *regardless of rank or status*. This is one of the few instances where a living member of the military will receive salutes from members of a higher rank. Therefore, LeMay was obligated to salute Doolittle.

O'Hollaren's dilemma was further compounded by the fact that Doolittle had been Lemay's *superior* 22 years earlier when he commanded the US Eighth Air Force in England as a Lieutenant General (3 stars) when LeMay was a Major General (2 stars) commanding the 3rd Bomb Division – part of Doolittle's Eighth Air Force as shown here.

Doolittle on the left (wearing 3 stars) with LeMay (wearing 2 stars) in England in 1944.

Doolittle was also 10 years older than LeMay. That compounded the problem, as seniority generally brings honor. On the other hand, the two had very different dispositions.

As the Directors continued to ponder the quandary, the focus shifted from military rank to the *contrasting personalities* of the two generals. LeMay was widely recognized as a tough, unbending, almost belligerent commander who expected his men to follow him. He had earlier told me, "I flew the lead aircraft on the first raid over Berlin." His nicknames included "Old Iron Pants", "The Demon", and "Bombs Away LeMay." His chomping on an ever-present cigar symbolized his no-nonsense personality.

Doolittle, though also an intense and focused personality, was personally affable and friendly. His smile was contagious. I mentioned to the Directors that I had also met him earlier and found him very approachable.

So on the basis of *personality* rather than *rank*, we hastily reached consensus that LeMay should receive the final plaque – even though strict protocol would have ruled that it be given to Doolittle.

As the presentation ceremony reached the time when I was to introduce Jimmy Doolittle – and whether or not due to the hasty decision the Board had just reached, I made a Freudian slip in describing Doolittle as "the extinguished – I mean *distinguished* — Jimmy Doolittle!"

Presenting a National Aviation Show plaque of honor to Jimmy Doolittle in Los Angeles — 17 May 1966

Jimmy good-naturedly joined the loud laughter at my *faux pas* as I presented his plaque to him. Many years later – in November 1983, I visited Jimmy in his Monterrey CA office where he answered – for an hour on tape — every question I had ever wanted to ask him about his colorful career… but my embarrassing 1966 introduction never arose.

The Opening Ceremony ended when General LeMay graciously and very formally received his plaque from me – as the most honored recipient.

Two additional occurrences related to this ceremony held personal significance for me.

First, my parents – visiting from Spokane, Washington — were able to attend the ceremony. Jimmy Doolittle and my father Wesley were born on the very same day: *14 December 1896*. I was honored to introduce them to each other.

Second, Jimmy Doolittle was promoted to 4-star General in 1985 by a special act of Congress, with President Reagan pinning those stars on him. So my wife and I attended his funeral and burial in Arlington National Cemetery on 1 October 1993 with all the Ruffles and Flourishes due a 4-star General – including the beautiful Air Force 4-Star General flag unfurled in a breeze (almost hiding that 4th star) behind the casket in as we approached the gravesite.

Presenting a National Aviation Show plaque of honor to Curtis E. Lemay in Los Angeles – 17 May 1966

Jimmy Doolittle's 4-star Air Force General flag at his burial in Arlington National Cemetery on 1 October 1993

What is the *significance* of recounting this concern about properly honoring two men? By what *standards* should honor be given any individual? At end of life, how can any person be *ranked* for honor? As death approaches for *anyone*, what importance should they place on being honored?

Signs for a New Century

When the 21st century dawned, seven billion people living on earth faced death. Not all would die at once, of course. However, as you read this book, *your* death is well in sight and unavoidable.

That record number of people (7 billion) yet to face death should stimulate increasing interest in many aspects of death – how to prevent or delay it, why it is inevitable, and whether it is the end of existence. There are abundant signs — if sought — that could indicate both the direction and dynamic that human death will exercise in this century.

Having personally known 10 of the 12 men who walked on the Moon in the 20th century, I can only speculate on what will transpire in space in the 21st century.

Navy Captain Edgar D. Mitchell, 6th man to walk on the Moon, unfurls an American flag 5 February 1971

The sixth man to leave his footprints on the Moon – Edgar D. Mitchell – and I met soon after he returned on Apollo 14. He and Alan Shepard spent over 33 hours walking on the Moon and brought back 100 pounds of Moon rocks when they splashed down in the South Pacific Ocean.

While Shepard made fame for hitting two golf balls on the Moon, Ed was better known for his interest in human consciousness. Ed and I had a three-hour discussion about his experiments between earth and the Moon with ESP (extrasensory perception — including reception of information not gained through the recognized physical senses but sensed with the mind) and its relevance to spiritual life.

As is often said, "truth is stranger than fiction" and it certainly applies to how Ed and I *happened* to meet. His mother, Ollidean Mitchell — residing in Tahlequah, Oklahoma – *happened* to receive a tape recording of a talk I *happened* to deliver sometime somewhere in which I *happened* to discuss my NASA involvement as well as my Christian faith. Some unknown person *happened* to mail that tape to her. She *happened* to listen to the tape – then *happened* to believe that I might somehow meet her son.

Somehow, she *happened* to locate my California home address. On 3 June 1971, she sent me a handwritten letter asking a simple question: "If the opportunity presents itself – God leading – will you please witness to my son to the Glory of God?"

Ed had returned from the Moon on 6 February – less than 4 months before I received her letter. Yet – in a series of incredibly rare circumstances – I was having breakfast with her son only 3 months later. *Opportunity did present itself!*

Apollo 14 astronaut Ed Mitchell having a 3-hour breakfast with me at the Ambassador Hotel in Los Angeles on 28 August 1971—less than 7 months after he walked on the Moon. We are contrasting his ESP views with spiritual warfare described in the Bible (II Corinthians 10:3-5)

Following the 2003 tragic re-entry disintegration of the Space Shuttle *Columbia* and deaths of seven astronauts, I was invited to deliver the keynote address on spaceflight safety at NASA's Marshall Space Flight Center. My address was a sober, realistic evaluation of both NASA successes and failures since its creation in 1958.

Those seven deaths on *Columbia* caused a two-year delay in US space flight – to hopefully remove factors that had led to the Shuttle destruction. Death is always the limiting determinant – the factor that decisively affects the nature even of space exploration.

Delivering the keynote address "Safety in NASA's Future" at the annual Safety Day at Marshall Space Flight Center on 27 October 2005

At the conclusion of my address, I was presented a mounted American flag that had flown aboard the Space Shuttle *Discovery* on the first rendezvous with and fly-around of Russia's space station *Mir*.

Personally, that flag is a treasured symbol of more than 40 years I was involved in space issues and NASA projects. Yet, does it symbolize any more than a frozen moment in time?

The Shuttle program has passed away – celebrated by *Discovery* now being displayed in the National Air & Space Museum at Dulles International Airport. Even America's singular leadership in space is no more.

American flag flown aboard Space Shuttle Discovery as it rendezvoused and circled the Russian space station Mir in February 1995 on a 9-day mission

What influence will human death play in space exploration in the 21th century?

Extend the *Trends*

There are a number of existing trends regarding the impact of death on humanity that have already been addressed – such as disguising it, denying it, deferring it, and defusing it. The most recent one, of course, is preparing for death by arranging the freezing of cadavers in anticipation of medical miracles to defeat its finality. There is little reason to doubt that all these trends will continue to exist.

The current trend of honoring people – particularly prior to their death as was illustrated in the National Aviation Show, will also undoubtedly be retained in this century. Honor, at best, is an abstract concept. It seems to combine a perceived quality of worthiness and respectability due to actions in accord with the moral code of society at large – but *at that time*.

Yet the honor results would be confusing indeed, were *all* the dead to be restored to a living state because the *honoring standards* over thousands of years have varied widely. Would there be open competition between *Moses* (1391-1271 BC), *Plato* (423-348 BC), and *Caesar* (100-44 BC) for the greatest honor? And what criteria would determine that distinction? Who would be the judge?

Questions as Keys to the Future

Another approach or source for signs regarding death in this century may occur by a process of inquiry – being aware of how society pursues answers to the seemingly unanswerable.

Death is a subject that has historically seemed to provoke many questions about "why?" Puzzles, perplexities, enigmas, and riddles seem to swirl about death – never answered.

Consider the works of Shakespeare, for example. *Romeo and Juliet, Julius Caesar,* and *Richard II* are permeated with poignant reminders of death without answers. This question from *Hamlet* (Act III, scene 1) is typical:

Who would fardels bear,
   To grunt and sweat under a weary life;
But that the dread of something after death,
   The undiscover'd country from whose bourn
No traveller returns, puzzles the will
   And makes us rather bear those ills we have
Than fly to others that we know not of?

The unresolved but fierce competition in America between Judeo-Christian and Secular Humanism worldviews also continues to raise serious thought and signs concerning death in 21st century society.

For brevity, these signs can be identified as *death questions*.

- Who will *define* it (physicians or politicians)?
- How much *authority* will be exercised over it (in fear of overpopulation or insufficient social security)?
- What influence will be used to *delay* it (by physicians and technologists)?

- What influence will be used to *accelerate* it (by politicians and sociologists)?
- To what degree will the continuing demise of the sacred, the excellent, the sublime impact it?
- What role will *suicide* — self-inflicted, physician-assisted, or socially-demanded (for population control) — play?
- How will *environmental concerns* (either as polluters or recipients of pollution) influence its timing?
- How could widespread refusal to tolerate *pain* expedite it?
- To what degree will it be accelerated by outbreak and proliferation (due to mass transportation) of *infectious diseases* with no medical cure?
- How will hedonism, abandonment of family structure, and unlimited sexual licentiousness influence it?

Conflict Between *Knowledge* and *Wisdom*

There is another source of indicators or signs relevant to death in the 21st century. The cultures, societies, nations and collective groupings of humanity continue to contribute many exhibits of death's influence.

Many view history as progressive (but not necessarily wise, good, or positive). Its record of what has transpired concerning death during the first 20 centuries may suggest a foundation for the 21st century. Without doubt, *information* has increased — aided and abetted by technology that enables its widespread dissemination.

However, the historic relationship between information, knowledge, and wisdom is incoherent at best. Seemingly, these three should be sequentially linked – leading to a positive outcome or "lessons learned" concerning death. Yet increasingly violent wars have raged throughout history – slaughtering millions indiscriminately. War looms constantly on the horizon as inevitable – even today.

T.S. Eliot – 20th century essayist, publisher, playwright, literary and social critic, as well as one of its major poets, once observed:

"Where is the wisdom we have lost in knowledge? Where is the knowledge we have lost in information?"

Several conclusions can be drawn from Eliot's observation:
- Information, knowledge and wisdom are related to each other.
- All three involve thought, conclusions, and action
- They have *progressive* order – from information to knowledge to wisdom

Information is the communication or reception of knowledge or intelligence – something transmitted or exchanged.

Knowledge understands something — especially about a particular subject, having awareness of facts and truths, or something that can be known. In other words, knowledge is information of which someone is aware. Knowledge can also mean the confident understanding of a subject, potentially with the ability to use it for a specific purpose.

Wisdom is the ability to use knowledge and experience intelligently – having the capability of determining what is wise versus what is unwise.

While knowledge is gathered from learning and education, most agree that wisdom is gathered from day-to-day experiences and is a state of being wise. Knowledge is merely having clarity of facts and truths, while wisdom is the practical ability to make consistently good decisions in life.

Lord William Thomson Kelvin, the Scottish mathematician and physicist known for his self-confidence once said:

> "When you can measure what you are speaking about, and express it in numbers, you know something about it; but when you cannot measure it, *when you cannot express it in numbers, your knowledge is of a meager and unsatisfactory kind.*"

Lord Kelvin surely knew better. Can you imagine him having *numerical expressions* for fear, love, beauty, ecstasy and toothache? He must have lived a very dull life.

However, the idea of having *numbers for everything* has seemingly caught on with many. Numbers are so clean – no caveats, reservations or qualifiers. Computers gobble them up and spit them out instantaneously.

In my earlier-mentioned NASA keynote address, I created and defined the term "datacide" – the fatal disease in Figure 1 of being overwhelmed with numerical data intended to make tough qualitative decisions easy.

> **DATACIDE**
>
> **DATA** –that *may* provide
> **INFORMATION** – that *may* provide
> **KNOWLEDGE** – that *may* provide
> **WISDOM** – data's only useful product
>
> **BEWARE!** *"Dazzling Display of Deceptive Decimals!"*

Figure 1 — *Datacide — the product of quantification run amok – using numbers in a frenzied, out-of-control, or unrestrained manner.*

What I have often called *"the dazzling display of deceptive decimals"* is employed to justify conclusions needing to be exposed and recognized for their fallacy. Yet, because when obvious foolishness is described by *numbers*, it passes the test for accuracy and authenticity.

It is as though wisdom is no longer needed. But humanity suffers for its loss. And death remains just as unexamined in Century 21 as it has for the previous 20 centuries. Wisdom beckons wistfully from the sidelines . . .

Political correctness in the 21st century has led to abject and pathetic denial of the undeniable . . . inventing language intended to hide the obvious. And it bespeaks a frightful society devoid of reality and willing to believe lies. This is beyond *lacking* wisdom . . . it is *running away from it!*

The Role of *Self*-Discipline

James Allen, in his book *As A Man Thinketh* [1], said:

> "Circumstance does not make the man; it reveals him to himself. No such conditions can exist as descending into vice and its attendant sufferings apart from vicious inclinations, or ascending into virtue and its pure happiness without the continued cultivation of virtuous aspirations."

That insight invites each one of us to recognize and weigh the significance of *motivation* — the private, internal reasons one has for acting or behaving in a particular way. Concerning death, everyone likely acknowledges, avoids, or anticipates it differently.

Despite the 21st century's exponential increase in number of human beings, the role of the individual has not diminished. Death means just as much today to each person, as it has ever meant in the past. Death is *individualistic*. Therefore, everyone faces death based on factors, situations, values, environments, and conditions unique to them.

Pain as Determinant

Whether we recognize it or not, pain and death are related. Pain is an unpleasant sensory and emotional experience associated with actual or potential tissue damage. It is a likely warning that normal health is threatened. But how long does "normal health" last? Ruling out accidents, natural disasters, war, and plagues, normal health ends at death. There is no storybook "living happily *ever after*."

So pain is avoided whenever and however possible. Pain-killers of all types abound – many producing addictions. Yet death awaits everyone as the ultimate pain . . .

Collapse of Historic *Moral* Values

The 21st century is the recipient of a long history of attempts to identify, document, and espouse standards of good and honorable living. These established codes or accepted notions of right and wrong are the underpinning of civilization.

The United Nations was founded in 1945 on such high and noble aspirations as faith in human rights, dignity and worth of the human person, equal rights of everyone, justice and respect for all, and promotion of social progress. Who could argue with those ideals?

Yet check the record. Wars have continued, terrorism flourishes while massive millions are fleeing from all kinds of persecution, starvation, and barbarism. Billions are spent every year to resolve never-ending worldwide conflict — without success. Viewed from outer space, the scene is chaotic and growing worse. *Something is wrong.*

The Judeo-Christian foundation of America is fast crumbling under Secular Humanism's denial of death – taking with it abandonment of historic underpinning of such cornerstones as commitment to marriage vows, revered family structure, accountability to God, law-and-order, financial thrift, sanctity of life, and acknowledgement of evil.

Disappearance of Deferred Gratification

Self-centeredness – a cornerstone of Secular Humanism – continues to dramatically influence many 21st century values, especially in the Western world. Under the guise of redefined freedom (i.e., abandoned responsibility), a new value system is emerging. This liberation from obligation has unleashed an era of "instantly having what you want when you want it."

The age-old requirement of waiting until one has the required resources before acquiring anything has disappeared. Credit cards assure that you can have any and every thing you want *right now*! And that includes housing, entertainment, automobile, vacation, clothing, food – even a sexual partner.

How is *death* related to this phenomenon? Did you ever hear the old saying, "Eat, drink, and be merry – *for tomorrow we die?*"

Technology's Ubiquitous Influence

All the foregoing signs proposed as influencing the direction and dynamic that death will exert in the 21st century have been greatly accentuated by remarkable and wonderful products of technology. However, the *blessings* of technology possess the potential counter effect of *crippling* society (and resulting in massive death).

Inherited from the 20th century is the political threat of using nuclear weapons and EMP (electromagnetic pulse) devices to cripple life as we know it. An EMP event can occur either naturally (through solar flares) or artificially, as the result of a high-altitude nuclear explosion.

Critical infrastructure protection, commonly referred to as CIP, is a priority in America for the Federal government, as well as the private sector and state, local, and tribal governments.

Infrastructures are the complex and highly interdependent systems, networks, and assets that provide the services essential in daily life. As shown in

Figure 2 they are currently organized into 18 critical infrastructure and key resource sectors managed by the Department of Homeland Security against a host of risks – foreign and domestic.

Critical Infrastructures to be **PROTECTED**

- Banking and Finance
- Chemical
- Commercial Facilities
- Commercial Nuclear Reactors, Materials, and Waste
- Dams
- Defense Industrial Base
- Drinking Water and Wastewater Treatment Systems
- Emergency Services
- Energy
- Food and Agriculture
- Government Facilities
- Information Technology
- National Monuments and Icons
- Postal and Shipping
- Public Health and Healthcare
- Solid Waste Management
- Telecommunications
- Transportation Systems

Figure 2 – *Critical Infrastructures to be PROTECTED*

Technology has been a major contributor to all those 18 sectors of government concern – in both positive and negative ways. For every *benefit*, there has been a corresponding *downside*. Technology offers no "free ride." In fact, every one of those 18 seemingly contributory enterprises, activities, and functions can easily become sources of *massive death*. It is that potentiality that continues to force expenditure of millions of dollars to avoid those risks.

Massive, Impersonal Death

The remarkable explosion of world-wide travel – enabled by airlines since World War II – has generated myriads of benefits. Cross-cultural communication in language, dress, customs, and even foods has erased many barriers as a result. Historic events are frequently viewed in real time across the globe.

On the other hand, technology has encouraged concentration of population through high-rise construction techniques that, in turn, provide opportunities of high vulnerability to earthquake, tsunami, and other disasters. Airliners have rapidly increased in payload capacity so that over 500 people can travel together at near-supersonic velocity for thousands of miles. Such remarkable achievements, of course, raise expectations of safety that are dashed when an airliner crashes and disappears!

Among hundreds of TV interviews I have given through many years, those associated with Malaysian Flight 370 from Kuala Lumpur, Malaysia to Beijing that disappeared in March 2014 demonstrate the international intrigue over unresolved deaths. Two years after its mysterious disappearance, I continued to be called for interviews concerning that mysterious disappearance!

One single *Sky News Arabia* interview I gave on that crash was simultaneously – and astoundingly — viewed by 70,000,000 in the Middle East! Could it be that 1% of the entire world population of 7 billion watched that interview?

One of my interviews on Mystery 370 — an ongoing TV series featuring enticing speculation on what happened to Malaysian Flight 370.

Role of Social Media

Perhaps the most visible phenomenon attributable to technology that began to appear in 2004 are forms of electronic communication — known as Web sites for social networking and microblogging – through which users create online communities to share information, ideas, personal messages, and other content such as videos.

On the personal level, the explosion of social media could only happen with technology's creation of the remarkable phenomenon of cloud computing — a model for enabling omnipresent network access to a shared pool of configurable computing resources. The "cloud" has changed so much so rapidly. Availability of high-capacity networks, low-cost computers and storage devices as well as the widespread adoption of hardware virtualization, service-oriented architecture, and autonomic and utility computing have led to a revolution in human interaction. Telephones and even personal computers can be bypassed by instantaneous communication.

Yet what about the cloud's downside? Sacrificed is personal privacy, stolen identity, enhanced international espionage, diverted attention (e.g., texting while driving), even loss of person-to-person interaction (everyone staring at a hand-held device instead of looking at another person).

Crime's sophistication continues to parallel and match the advances of technology. And thereby *the threat of death increases.*

Promises That Depend on *Decisions*

The alluring side of technology includes continuing reduction — through mechanization — of personal labor required to obtain so many desirable benefits: world-wide travel, access to unlimited sources of information, controlled living environment, clean water, variety of safe and healthful foods, and unrestricted communication.

Those technological wonders are realized by decisions we all exercise – knowingly or subliminally – to accept those promises. As incredible as it might seem, all those promises are vulnerable to being *worthless* — should inadvertent or deliberate forces to eliminate them occur. That possibility is what worries those charged with CIP.

Imagine what would happen should GPS tracking be knocked out via EMP while 5,500 airliners over the US, on any typical day, are in the air. The pilots would instantly be unable to know where they were, how to navigate to their destination, and what other aircraft were in their path. *Thousands of deaths would be unavoidable . . .*

Extension of Life *Duration*

People will apparently continue to live longer lives in the 21st century. Secular Humanism's only domain — "time between birth and death" – will be extended as a result of technology. The quality of that lengthened time will obviously vary due to (a) availability of technology's benefits and (b) protection against the unthinkably devastating products of technology like nuclear weaponry and bacteriological warfare.

Increasing use of drugs (both legal and illegal) to enable people to become "other than they are" will likely influence the *quality* of extended life. Societal values tend to be related to the degree that this modification – and its correlated behavior – can be tolerated. People when thus modified are seldom viewed as productive citizens and often become a burden to be tolerated. Technology will continue to produce both benefit and loss for humanity – the latter being *premature death.*

Revision of Life's *Values*

Throughout history, humanity has generally reflected the significance, quality, and importance of life. Individually and collectively, life appears to be more than random happenstance. Yet its *existence* has caused many to wonder about it.

Some, like renowned French mathematician, physicist, inventor and writer Blaise Pascal (1623-1662), have postulated questions about it:

> "When I consider the short duration of my life, swallowed up in the eternity that lies before and after it, when I consider the little space I fill and I see, engulfed in the infinite immensity of spaces of which I am ignorant, and which know me not, I rest frightened, and astonished, for there is no reason why I should be here rather than there. Why now rather than then? Who has put me here? By whose order and direction have this place and time been ascribed to me?"

Each generation seems to acknowledge *values* – significance, meaning, implication, and import – well beyond simply existing. Yet those values change over time. So values are dynamic, not static. Technology — loaded with promise — provides a driving force in changing values and will continue to do so in the 21st century. It is generally accompanied by a downside that, even in the 21st century, will not overcome the inevitable collective character of human death.

In a broad scope, Figure 3 depicts ten historic phases that humanity has seemed to follow for centuries. This "cycle of civilizations" would indicate that reversal or elimination of any of these civilizations is unlikely.

The Historic "CYCLE OF CIVILIZATIONS"
(Adapted from Arnold Toynbee)

- From *bondage* to *spiritual faith*
- From *spiritual faith* to *courage (to take risk)*
- From *courage* to *freedom*
- From *freedom* to *abundance*
- From *abundance* to *selfishness*
- From *selfishness* to *complacency*
- From *complacency* to *apathy*
- From *apathy* to *fear (of taking risk)*
- From *fear* to *dependency*
- From *dependency* back again to *bondage*

Figure 3–*The Historic "CYCLE OF CIVILIZATIONS"*

In which civilization would *you* place society today?

Chapter 12

Managing Your Risk of Death

"Happy is the man who has broken the chains which hurt the mind, and has given up worrying once and for all."— Ovid

Just as I was leaving my Washington DC hotel room one morning in March 1974, the phone rang. I stopped and returned to answer it.

"Dr. Grose, I'm calling to see if you would be interested and available to serve on a board we are forming to address the risk of common-corridor operations between the new rapid rail transit system (later called METRO) and existing railroad traffic."

It was an executive of DeLeuw, Cather & Company – the general engineering contractor building the system – on the line.

"We're aware of your expertise in managing risk. If you agree to serve, I will confirm it by letter of invitation to our first meeting next week."

Thus began my involvement – on a 5-man Board of Consultants formed by the Washington Metropolitan Area Transit Authority (WMATA) under pressure from the Congress to identify, rank, and control many obvious but yet-unidentified risks.

WMATA had decided at the outset of this multi-billion dollar project — *without considering attendant risks* – that it would be more economical to route the METRO tracks in existing railroad rights-of-way than to condemn, buy, raze, and clear desired properties. Since METRO would primarily be a series of city-to-suburb linkages and there were already rail lines approaching Washington from all directions, the decision seemed a good way to save money.

However, those potential savings might easily be wiped away if certain credible accidents were to occur. Imagine – for a moment — spreading open historic railroad tracks that had been there for over a hundred years to insert new tracks between them for high-speed, lightweight commuter trains.

It was no surprise that this idea would be opposed by Chessie System, the primary railroad right-of-way owner. The prospective interruption of its operation to permit re-routing of tracks was not its only objection. More significantly, running its long heavy trains at minimum passing distance from relatively fragile commuter cars loaded with up to 800 people at rush hour with closing velocity of 135 mph (train at 60 mph, transit at 75 mph) portended a death toll exceeding the worst airliner crash!

So the initial meeting of the Board of Consultants was tense. Chessie's Chief Engineer, WMATA's Chief Engineer, a renowned retired railroad expert, and DeLeuw Cather's Executive Vice President were my four colleagues on the Board. *I* knew little about railroads while *they* collectively knew nothing about managing risk. Among all those railroad experts, I was definitely a minor player.

After a round of formal introductions, DeLeuw's Executive VP – as Chairman – opened the meeting with this startling comment: "Well, I have no idea *why* we are here, but we evidently have an assignment or obligation concerning risk."

I was shocked... but remained silent.

Slowly and awkwardly, some began offering brief thoughts about available railroad data on load shifting, derailment, switching, speed control, signaling and braking that might be relevant. Hesitant discussion haltingly headed into virgin territory for everyone. There was no agenda. The meeting drifted along in random fashion for an embarrassingly long time.

Finally, I proposed that – since *risk* was of concern – there was a *systematic* approach to managing it.

That seemed to bring immediate relief to the almost meaningless palaver about possible railroad problems. Since I was teaching courses on systematic management of risk at nearby George Washington University, they readily agreed to my offer to lead the Board — by integrating its extensive railroad expertise — in applying the systems approach to WMATA risk.

From that point forward, my offer provided the framework for the Board's work. Over many months, the Board focused its expertise and imagination on identifying possible WMATA-Chessie risks that were then evaluated and ranked for significance. That ranking, in turn, became the foundation for selecting and investing in common corridor risk countermeasures.

Risk Must Be *Identified*

No risk can be *managed* unless it is first *identified*. The initial inclination of the railroad experts on the Board had been to recall general *solutions* for avoiding losses due to risk; e.g., how to evade derailment or improve braking. This was analogous to discussing airlines, airliners, and airports with no particular destination in mind.

Obviously, the key to managing METRO risks was to identify them *before* they could happen. But how could the Board do this? Was there a crystal ball to aid in clairvoyance? Would this mean random brainstorming or throwing darts while blindfolded? Was there a special ability to predict what will happen in the future?

The answer I offered consisted of mobilizing the expertise of the Board for three tasks:

- Create a functional model of WMATA as a *system* – with its known inputs and desired outputs – defining sequentially step-by-step *what* must happen to convert those inputs into outputs.

- Using the WMATA model as a stimulus, conceive for every function three short stories known as *risk scenarios* that described how a specific loss could occur – what, who, when, and how large a loss would result.

- Create three alternative *countermeasures* for every risk scenario, estimating the cost of each one.

The Board then used this process to create and rank 210 risk scenarios – assigning values for severity, likelihood, and estimated cost of prevention to each one. This imaginative work required lots of time, discussion, and research. And it wasn't all accomplished in an office environment. The five

Board Members all toured Chessie tracks in an automobile equipped for riding on rails — visible behind us — to obtain a physical perspective that matched our theoretical views and conclusions.

Washington DC METRO Board Members touring Chessie-METRO common corridor to review potential risks-

When assembled into a ranked array (known as a *Risk Totem Pole*™), the Board's work proved to be a remarkable and systematic response to Congressional concern when it was presented in a press conference.

It was obvious that WMATA had not only understood the breadth and severity of risk involved in common corridor operations. It had even authorized the *cost* to manage the risks it had identified — over $25,000,000! An admirable beginning...

Identified Risks Require *Implemented Action*

Good beginnings do not guarantee good outcomes. When WMATA began implementing countermeasures associated with those risks having the highest severity and probability at the top of its Risk Totem Pole™ – the ones with the "biggest bang for the buck" in reducing risk, they ran into headwinds from parties who opposed them.

That should have been anticipated because whenever change in the status quo is proposed, there will likely be opposition. However, the most significant WMATA risks involved the possibility of either a METRO or Chessie train penetrating the opposing track's domain to result in an incredibly deadly collision.

To prevent or at least minimize the magnitude of such a disaster, it was imperative that either (a) an automatic signal be sent to operators of both trains that their domain had been violated or (b) that operators be able to contact and warn each other by radio to immediately stop in an emergency. Many of the countermeasures for METRO risk scenarios required direct METRO-to-Chessie train communication. But *this critical prevention has never been implemented.*

I have even testified before the Maryland State Assembly that radio contact between the two trains was essential for safe common-corridor operation — to no avail.

To this day, there is no direct contact between METRO and Chessie train operators, and environmental interests have continued to defeat the only alternative. My continuing concern about this high risk is affirmed in my *Washington Post* editorial in Figure 1.

Personal Risk of Death

What relevance does WMATA's management of *its* risk have to your management and control of *your* most critical risk – *death*?

Believe it or not, there is a correlation. And it is worthwhile to learn some risk lessons from WMATA – because the stakes are far higher for *you* personally than those faced by WMATA!

As shown in Figure 2, both WMATA and you have risks to manage. Those risks can be managed effectively or poorly – reactively or with foresight. Your personal risk of death, however, far exceeds all others – including WMATA's — in *significance*. Why? Because, as you recall from Chapter 8, you face not *one* death but *two*! Remember: *everyone dies twice.*

Managing Your Risk of Death

Figure 1–*WASHINGTON POST Editorial* – 19 June 1987

COMPARATIVE MANAGEMENT OF RISK

WMATA	YOU
• Impersonal	• Personal
• Temporary	• Permanent
• Incomplete	• Final
• Vulnerable	• Established
• Political	• Spiritual

Figure 2 – *Comparative Management of Risk*

The title for this book intentionally declares that "death is not *fatal.*" More than "a play on words" — a turn of phrase with a double meaning, there is definite purpose in correcting a popular misconception about the finality of death as "the end of everything."

In common parlance of accident investigators, the number of people killed is frequently referred to as "fatalities" or simply "fatals." Yet "fatal" actually means "the development of events beyond a person's control and determined by a supernatural power." You may have heard statements like *"Fate decided his course for him"* or *"Her injury is a cruel* **twist of fate.***"*

In Greek and Roman mythology, each person's destiny was thought of as a *thread* spun, measured, and cut by three Fates: *Clotho, Lachesis,* and *Atropos.* The reason *fatal* was selected as part of this book's title was to deliberately reveal the superstition, deception and fallacy it attempts to hide.

So – with cleared understanding about managing the risk of YOUR death, let us lay a sure foundation. It begins with the *systems approach* that was defined in Chapter 1. The significant role of *risk* was elaborated in Chapter 2. Now, in this final chapter, we are ready to not only combine and focus these two concepts on the topic of your death but *personalize* them as well. After all, *your* death is vitally important to *you!*

Management's Functional and Volitional Elements

It may appear a bit radical to view the risk of your own death as being *self-managed*. Of course, suicide is ruled out as an option — neither considered nor implied. However, managing *anything* connotes planning, organizing, and controlling it. And motivation is a vital part of management as well.

What role can *planning* play in managing your personal risk of death? It implies getting ready, making certain, and assuring availability of everything required. Most importantly, *foresight* is required – the application of wisdom, anticipation, expectation, and results. It defeats and takes the sting out of surprise.

To *organize* your own death's risk involves "bringing order out of chaos" wherever and whenever possible. Deliberate commitments to prepare...and to avoid randomness are indicators of organization. *Control* is the resulting follow-through action of organization.

Opportunity and *Incentive* Are There

The inevitability of human death provides — even invites — both *opportunity* and *incentive* to manage its risk. Obituaries are constant reminders that your departure from this life lies ahead. Have you ever pondered what yours will contain? What intrigues you most about obituaries – especially those that summarize the life of someone whom you've known intimately? How important do you believe your own documented life story will be?

Open and thoughtful recognition of your impending death provides you with an ample foundation for managing its risk. Ironically, far more attention is likely paid to preparing a backward-looking will than to managing one's destination after death. This lop-sidedness is increasingly true due to Secular Humanism's refusal to acknowledge death.

Goal – Not Guarantee

As with all managerial endeavors, managing your personal risk of death assures no identifiable *guarantee* of its effectiveness because it is an ongoing activity until you die. Further, it is impacted by distractions, timing and lack of information. But it is – and should be — the goal of everyone to manage his or her risk of death with foresight and conscientious motivation.

The "risk of death" being managed obviously involves – not only the inevitable physical death that is universal but also the impact or result following that event. This includes resolving whether physical death is the end of existence or whether life continues – where, how and in what venue.

Likelihood is Fixed — But *Magnitude* is Yours

Two dimensions of human risk – its *likelihood* combined with its potential *magnitude* – are discussed in Chapter 2. Yet concerning your personal risk of death, only the second dimension is available for your management — since none of us avoid the first's inevitability.

Perhaps you recall in Chapter 6 my co-worker Arnold's dramatic retort "What do you want to do – live *forever*?" Whether or not you desire or intend it – you *will* live forever! Therefore, human risk's likelihood is fixed at 100%.

But what is meant by managing the "*magnitude* of the risk of your death?" What can you possibly manage – given the potential immensity, greatness, extent, grandeur, and unlimited vastness of the *unknown* beyond death – to reduce that risk? Does magnitude imply *dimensions*? Is death the same for *everyone*?

You need not be intimidated or overwhelmed to conclude that you *can* manage the magnitude of your risk of death. In fact, though it appears to be a *limitless* risk due to many unknowns, it can be reduced to the simple decision to experience *moral* death prior to *physical* death (see Figure 3 in Chapter 8).

It is worth noting that, if Secular Humanism is your worldview, you have already decided that *moral and physical death do not happen* since they are never even acknowledged.

Objective: What *Type* of Death?

It is rare for anyone to openly discuss dying or death – particularly their own. Yet there is nothing as certain and unavoidable. Perhaps it's because we like to be positive, and death seems to be the epitome of negative.

Wisdom, on the other hand, suggests that *everything* that influences our death should be considered. At the outset, an attempt to define what *type* of death one wishes to die could be critical. This might mean considering the desired *conditions, timing, location,* and *mechanism* of death – at least those within your ability to control.

However, before establishing all those criteria for *your* death, it is essential that the prevalent idea of human death that was bounded and defined as a *system* in Chapter 1 be examined further. In Chapter 8, the premise of two *types* of death was introduced – with its corollary that everyone dies twice. And you can *definitely control* the first death!

Perhaps the oldest written document mentioning death is the Biblical book of Genesis. In that account, human life was initiated with a clear warning that death would result if the fruit of a certain tree was eaten. That fruit itself was not poisonous. Instead – and much more significantly, eating it violated a warm, loving, ongoing *relationship*.

When Adam ignored and violated that warning, he did not *immediately* die but lived for hundreds of years afterward. However, initiated instantly were *two types of death* that subsequently impacted the entire human race as a result of Adam's deliberate violation:

- **Death One**: The relationship between God and humanity was severed. Since that event, mankind has run away from God – hiding and attempting to cover its rebellion to remain independent of God and deny His existence. This is known as *spiritual* death – also defined as *moral* death in Chapter 8.
- **Death Two**: Human lifespan was no longer unlimited. Bodies have ever since "returned to the dust" from which they were formed. This is recognized as *physical* death.

It would seem obvious that the *primary criterion to be resolved prior to your death* is deciding whether you recognize and resolve Death One — since that must take place prior to Death Two (when your heart stops beating).

It's a vital matter of proper *sequence*. Consequences of wrong order are both *severe and eternal*.

Setting Criteria for *Your* Death

You may not often pause to ponder about your own upcoming death. Sometimes, it may rise to your awareness when someone close to you dies, when you attend a funeral, or when you read an obituary of a dear friend. On the other hand, the very idea of establishing specific standards, rules, or prescriptions regarding your death might seem abhorrent or too self-centered.

Why not just die *however and whenever...*

America — caught up in conflict between Judeo-Christian and Secular Humanism worldviews — is also greatly influenced by science and its stepchild *scientism* that was discussed in Chapter 7. Among the promises of science is that it might even hold the key to ultimately *overcoming* death because it has remarkably conquered so many diseases and thereby extended the span of life. That's the current charm, for example, of supporting stem-cell research.

However, Will Durant, Pulitzer Prize winning historian and philosopher, expresses in Figure 3 the conclusion that the contribution of science to the meaning and worth of life ends up being *cancelled* by death.[1] Why do you think that Durant reached that conclusion? Do you agree with him?

LIMITS of SCIENCE

"Science gives a man ever greater **powers** but ever less **significance**; it improves his tools and **neglects his purposes**; it is silent on ultimate origins, values, and aims; it gives life and history no meaning or worth that is not **cancelled by death** or omnivorous time."

Will Durant in *The Story of Civilization*

Figure 3 — *LIMITS of SCIENCE*

When you die, there is much more at stake than you might have imagined. Apart from the details of burial or distribution of property, the fundamental issue discussed earlier is summarized in a simple question: *What happens to YOU when your heart stops beating?*

It all depends on whether that event is your *first* death or your *second* death! You alone control that sequence.

Carrying Out *Your* Intentions

Many of us have engaged legal counsel to prepare a written will intended to carry out our intentions about inheritance values, relationships, and distribution of physical assets after our departure from this earth. Legendary stories abound of *failure* of wills to be executed as intended. Historian James Bryce, in his 1888 work *The American Commonwealth* said: "Jefferson might turn in his grave if he knew" — coining a phrase often heard to describe the disparity between *intentions* and subsequent *reality* of the deceased.

As we realize, we are powerless to assure that our intentions – after death – are executed as desired. It would actually be an odd world if that *did* happen. Think what it would mean if the post-death intent and desire of those who died 200 years ago prevailed today. What would transportation look like? How would world-wide commerce be accomplished?

Obviously, our intentions are *extremely* limited – based inevitably on our very circumscribed experience and view of reality. Most intentions involve interaction among personalities —- all of whom lie outside our sphere of influence once we depart this life. Beyond that limitation, there is considerable ignorance about how to even realize our intentions while we are still on earth – despite the benefits of education, experience, and motivation.

Bottom line: you are likely to have neither authority nor capability of being certain that your post-death intentions are accomplished.

Societal Support: Identifying and Assuring It

Moving from individuality to collectivism, what are the *external* forces and influences that impact your ability to manage the risk of your death? Some will be within your personal control. But many will reside where you are swept up in values, situations, and milieu well beyond any control you could possibly exercise.

Sustaining the *quality* of life — at both personal and societal levels – until your death is a foundational objective. This means that — beyond the *physical* aspects that preserve and extend bodily life like genetics, diet, exercise, sleep, and medical assistance — management of both the *intellectual* and *spiritual* dimensions of life deserve equal consideration for your possible management.

Obviously, societies around the world vary widely regarding family life, education, governance, housing, rule of law, health care, and private freedom. So the primary focus for controlling the quality of life that determines your risk of death will be on *personal* rather than *societal* values. What *you believe* determines how well you are able to manage your risk of death in whatever society you live.

Maintaining Your *Self-Identity*

Foundational to managing the risk of your death is determining who you really *are*. For centuries, philosophers have wrestled with answering the three questions in Figure 4. They have presumed that *everyone* – down deep – possesses an identity, awareness, realization, and cognition of their unique existence and destiny.

Three Ageless Questions

WHO *am I?*
WHY am I *here?*
WHERE am I *going?*

Figure 4– *Three Ageless Questions*

The idea of *self-identity* was pioneered in the mid-20th century by American psychologists Carl Rogers and Abraham Maslow – the latter also known for creating a "hierarchy of needs" predicated on fulfilling innate human needs in priority that culminates in *self-actualization*. Secular Humanism is focused exclusively, of course, on the "self" because it is considered the "be-all and end-all" of existence.

Regardless of your worldview however, there is a vital reason to pursue answers to those three questions in Figure 4: *ultimate accountability* that none of us can escape. We are not only *responsible* for how we live. We determine what happens to us *after death*! So it is critical that our self-identity remains intact and is not allowed to atrophy.

When and how does self-identity *form*? How do you *maintain* your self-identity? What is *involved* in doing so? Why is it *vulnerable*?

The *Obscure* but *Critical* Forces

Self-identity is dynamic — ever-changing to match the life you are living on earth. We all likely look backward in amazement at what we have experienced in life – because so little was due to forethought personal choices. We were caught up in a whirlpool of change from the moment of birth. Health, education, location, politics, housing, government, transportation, communication, friendships, and commitments frequently occurred without our weighing options or being aware of their consequences. We inherited them.

Those forces are still operative today in your life. Are you *maintaining* your self-identity or is it overrun by circumstance? Bringing one's *mind* and *soul* (the "self-identity") into alignment with the body will produce the optimum process leading to one's predetermined and desired death objective of managing risk of death.

Balancing the *Communal* with the *Personal*

Obviously, no one can override the forces and factors *outside* their control. But they must be balanced somehow with individual values for which we are responsible and determine our character. Typically, that balance involves interacting with social, political, religious, economic, legal, and technological issues that produce conflict requiring resolution.

Often this process of weighing, equalizing, and trading is done *subconsciously* with results that can be surprising when they become evident after a long period of unchallenged accommodation. Only then do we realize that we may have made undesirable long-term choices that slipped beneath our conscious awareness. Collectively, they form our *private* values – determining what we consider right or wrong, proper or improper, expedient or enduring.

It is almost as though everyone needs a lodestar – something to serve as an inspiration, model, or guide – to achieve the proper balance of values for managing their *personal* risk of death.

How are *you* aware of the balance you are maintaining?

MANAGING All the Way to the *Threshold*

Every one of us is on the same journey – one that ends in death. Hopefully, that common destination has been adequately defined, discussed, and focused. There is no ambiguity. Life on earth will end.

The overriding theme of this book is intended to be personal and focused – not abstract or ambiguous. So as we come to its conclusion – particularly focused on managing *your personal risk of death* as the reader, emphasis remains on continuity to completion. Whether the date of your departure from this life will be a surprise or proceeded by sufficient evidence of its arrival to allow final management decisions, prudence demands maintaining an always-ready state of readiness.

Approaching that threshold – the portal or gateway marking the end boundary of life's journey, we should recognize by now that there are *two options of departure* therefrom.

Life is often referred to as a journey. Regarding *optional* roads, Robert Frost's familiar poem *The Road Not Taken* is relevant as we consider approaching the threshold. Remember Frost's poignant conclusion "I took the one less traveled by, and that has made all the difference"?

There are two radically divergent roads leading toward that threshold. By now, it should be easy to acknowledge them since they are correlated to *Death One* and *Death Two* described in Figure 2 of Chapter 8:

- **Road One**: you voluntarily die *Death One* and await *Death Two*
- **Road Two**: you involuntarily die *Death Two* and await *Death One*

Everyone starts out on Road Two – because we are all born with inherited natures that are contrary to what the Creator intended at the outset of humanity for every human being. It is commonly called "human nature." Theologians describe it as a depravity or tendency to evil that is innate in humankind – transferred from Adam to all humans and thus called "original sin." It is the root of the self-centeredness promoted in Secular Humanism.

To move from Road Two to Road One is simple. But it requires a *deliberate commitment* based on recognition and acknowledgement of three factors:

- You are infected, bound and crippled by your human nature

- You desire to abandon and replace your human nature
- You accept the Creator's free offer of a *new* nature paid by the death and resurrection of Jesus Christ (this is known as *Death One*)

Recounting an interesting and relevant visit I once made may aid in illustrating how Death One is significant in moving to **Road One** – particularly related to the importance of electing it prior to experiencing Death Two.

Westbound on our way around the world, my wife and I departed Bombay (now Mumbai) India on a long flight late one November night. Our route took us northwest over Pakistan and Iran — passing over Tehran before turning west over Syria to the Mediterranean Sea.

For hours, our unbroken field of vision that night — as we overflew the oil fields of Iran — was filled with literally hundreds of bright flares burning off waste or unusable natural gas. It was something that I had never previously realized – the *deliberate wasting* of irreplaceable natural resources that must take place in order to obtain a fuel for internal combustion engines of so many types. That realization caused me to ponder anew the issue of *unlimited consumption of non-renewable resources* – especially fossil fuels whose rate of formation has required eons of time.

Our primary destination for this overnight flight was Tel Aviv, Israel. However, the captain informed us — after we were airborne — that landing in Israel was questionable because the Yom Kippur War was just winding down. The UN had brokered one cease-fire that had soon collapsed. Another one had been arbitrated but was still tentative.

So we flew on, uncertain about our destination...

It was dawn when we were amazingly cleared to be one of the first commercial airliners to land at Lod Airport in Tel Aviv! The second and final cease-fire had terminated the Yom Kippur War! Giant USAF C5 *Galaxy* transports were still landing however at short intervals, their war materiel continuing to be rapidly unloaded.

At this same airport the previous year, three terrorists — members of the Japanese Red Army – had carried out an attack in the massacre of 26 that also injured 80. So the high state of Lod Airport security that day was the highest I have ever witnessed!

DEATH Is Not FATAL

Our visit to Israel was noteworthy in so many respects. However, there was one event that surpassed all others in my entire life for its symbolic significance. Its location as well as physical form was overwhelmingly powerful to me.

It happened in a rock-cut garden tomb outside the north wall of Jerusalem and adjacent to the rocky escarpment considered by scholars to be Golgotha upon which Jesus Christ was crucified. That site is described in the Bible (in Matthew 27, Mark 15, and John 19).

I stood outside that ancient tomb wherein the body of Jesus Christ is recognized to have been laid following his brutal crucifixion over 2,000 years ago. Why did I consider it to be so overpoweringly significant to me?

It was empty!

The most significant event of my life – to stand outside the entrance to the only empty tomb in the world — in Jerusalem — and then enter it! Note my left foot in the channel for the stone that was rolled away in Christ's resurrection.

Certainly I had known about Christ's resurrection from childhood.

Easter – in the United States – is so widely celebrated in so many ways that it was inevitable that I would hear about it. Easter parades date back

194

to the Dark Ages. Since 1878, there has been an annual Easter egg roll at the White House. Great music like Handel's *Messiah*, new clothes, even the Easter bunny are all celebrated annually in honor of that empty tomb.

But I then *entered* that tomb, sat down, and began to speculate.

There are innumerable reasons – historical, political, cultural, and personal – that *the resurrection of Jesus Christ* overwhelms history. The primary and obvious one is to recognize that the Western calendar is divided in half – BC (before Christ) and AD (Latin *anno Domini* meaning "in the year of Our Lord Jesus Christ")

But most controversial is whether His body remained in the grave, was stolen to be buried elsewhere, or was resurrected in a new form. My left foot stood in the narrow inclined trough wherein a large stone had been rolled downward to seal the entrance after Christ's body had been wrapped and laid inside.

I sat *alone* in that empty tomb – for quite some time, pondering what had transpired 2,000 years earlier. Was it *noisy* when the stone rolled *backward* up that trough? What did the *transformation* of His brutally battered body into one that readily violated physical laws look like as it happened? How did the soldiers guarding the tomb react to this sudden, unexpected event?

You likely might have proposed and pondered many other different questions — had you been there with me. Without any doubt, that setting had to have been the site of an incredible, mind-blowing occurrence!

However, the reason for sharing this private 1973 experience is that *it established the fundamental difference between Road One and Road Two* – and thereby provides an announcement or declaration that every person now faces a critical decision.

The resurrection of Jesus Christ forces everyone to make a rational selection on which road to take before you depart life on earth — because it is the basis for all of us to die Death One!

That differentiation may be so subtle that it can slip by unnoticed. But it is very significant – *eternally* so!

Recalling Will Rogers' observation, "All men are *ignorant*, except in certain subjects," are you possibly ignorant of this *required decision*? Easter takes on a radically different meaning than Easter eggs, Easter clothes, Easter music, or Easter egg rolls! It announces that every person now has the option of *where* they will spend life after death.

Everyone today – as we all approach that threshold (the time of our death) – is either on Road One or Road Two. Road One is far superior to Road Two – assuring an infinitely superior destination.

Do you recall "*Debi's Dilemma*" in Chapter 1 when she was walking on Road Two – afraid to die? Contrast that situation with "*Debi's Resolution*" in Chapter 3 when she had moved onto Road One — free from fear of death!

Also in Chapter 1, I mentioned my desire to recount and share with you my own resolution of "getting ready to die" by moving from Road Two to Road One.

How did that happen for me? I actually died on 17 August 1945 in Spokane, Washington. It was my *moral* death, as I accepted Jesus Christ's death as a replacement for my deserved death – due to my inherited self-oriented nature, the same one we all share as human beings.

If you look once more at the Biblical Options Regarding Death (Figure 3 in Chapter 8), I exercised that green vertical option – "Believer's MORAL Death!" Just like Pat Boone's song says, I ". . . met the Son. He's the One!"

That option is available for YOU, too!

I hope that you will join Debi and me on Road One!
Remember Death is *NOT* Fatal!

NOTES

Chapter 1 — Human Death as a SYSTEM

1. Elisabeth Kubler-Ross, ***On Death and Dying***, (London: Routledge, Taylor & Francis, 1970)

2. Ernest Becker, ***The Denial of Death*** (New York: The Free Press, 1973), preface

Chapter 2 — The Meaning of Risk

1. Vernon L. Grose, ***MANAGING RISK: Systematic Management of Risk*** (Englewood Cliffs, NJ: Prentice Hall, 1987)

Chapter 3 — Human Obsession With Death

1. *The NIV Worship Bible* (Grand Rapids, MI: Zondervan, 2000)

2. Frederick J. Hacker, ***Crusaders, Criminals, Crazies*** (Toronto: George J. McLeod Limited, 1976), 7.

Chapter 5 — Demise of Noble Savage Theory

1. Marshall B. Rosenberg, ***Nonviolent Communication: A Language of Life*** (Encinitas, CA, PuddleDancer Press, 2008)

2. John Lukacs, ***At the End of An Age*** (`New Haven & London, 2002)

Chapter 6 — Modern Masking of Death

1. Sam Donaldson, ***Hold On, Mr. President!*** (New York: Random House, 1987)

2. Sherwin B. Nuland, ***How We Die*** (New York: Vintage Books, 1995), 243

3. Marie de Hennezel, ***Intimate Death*** (New York: Alfred A. Knopf, 1997), 26

4. Kathleen Dowling Singh, ***The Grace in Dying: A Message of Hope, Comfort, and Spiritual Transformation*** (New York: HarperCollins Publishers Inc.), 53

5. Ernest Becker, ***The Denial of Death*** (New York: The Free Press, 1973)

Chapter 7 — Death: Finality or Change of State?

1. Studs Terkel, ***Will the Circle Be Unbroken?*** (New York: The Free Press, 2001)

Chapter 8 — One Answer From Antiquity

1. William Strauss and Neil Howe, ***The Fourth Turning: An American Prophecy*** (New York: Broadway Books, 1997), 21

2. Paul Tournier, ***The Seasons of Life*** (Richmond, VA: John Knox Press, 1964)

3. Hiroshi Obayashi, ***DEATH AND AFTERLIFE*** (New York: PRAEGER, 1992), 112

4. ***The NIV Worship Bible*** (Grand Rapids, MI: Zondervan, 2000), II Peter 3:9

Chapter 10 — Toward a Rational Forum

1. Sherwin B. Nuland, ***How We Die*** (New York: Vintage Books, 1995), 10

2. Mike Huckabee, ***Kids Who Kill*** (Nashville: Broadman & Holman Publishers, 1998)

3. Os Guinness, ***The Dust of Death*** (Wheaton, Ill.: Crossway Books, 1994), 35

Chapter 11 — Death in the 21st Century

1. James Allen, ***As a Man Thinketh*** (The Savoy Publishing Company, 1903)

Chapter 12 — Managing Your Risk of Death

1. Will Durant, ***The Reformation*** (New York: Simon and Schuster, 1957), 4

Appendix A

Characteristics And Peculiar Laws Of LIFE'S FOUR SEASONS

The following conclusions have been adapted and summarized from *The Seasons of Life* by Paul Tournier, MD (John Knox Press, 1964). They may not necessarily be found in exactly the same format in the text, but their essence can be traced to thoughts in this remarkable and insightful book. Relevant pages of the text are shown in parentheses.

These descriptors for the four seasons provide the reader with a governing framework or infrastructure for classifying topics in ***DEATH Is Not FATAL***.

SPRING

- Life force is unleashed at birth — libido then propelling psychic mechanisms (11)
- Birth is an integral part of nature — from which neither progress in knowledge nor boldness of faith can free us (11)
- Belonging to two worlds at the same time — natural and supernatural — with both realities fused into a single life (11, 13)
- Need for love, protection, and tender care (25)
- Dependence on parents (21)
- Passive submission — concern for the forbidden (36)

- An age of legalism (36)
- Life is viewed in terms of absolutes (22)
- Early acquisition of a sense of destiny — an inner awareness of life having *purpose* — while being formed and unfolded by each event of life (19)
- Discovery of perpetual "becoming" in its infinite complexity (18)

SUMMER

- Moral self-direction — dependence on personal awareness of psychic controlling mechanisms (21)
- Governed by law of *action* (37)
- Adult action fed by carry-over of youthful idealism (23)
- Fullness of adulthood — realized by *becoming* an adult (21, 22, 32)
- Development based on 4 factors: love, suffering, identification, and adaptation (25-28)
- Establishment of home and career (37)
- Universal necessity to succeed (38)
- Increasing need for meditation (44)
- Intuition of a life plan whose accomplishment will mean life's fulfillment (40)
- Ever-expanding range of objectives to be attained (45)
- Realization of evil being inextricably mixed with good (22)
- Drawn by continuous change to awareness of ultimate destiny (19, 22)

AUTUMN

- Recognition that desired goals may not be realized (48)
- Gradual realization how inevitably *incomplete* life is (46)
- Wisdom derived through *integration* of lived experience (23)
- Learning to go back downhill — after long upward journey (47)

- Revision of values and priorities (51)
- Acceptance of one's age (52)
- Viewing of time as a diminishing capital (48, 53)
- *Doing* and *having* giving way to *being* (54)
- Searching for the *true* meaning of life (52, 55, 57)

WINTER

- Looking *ahead* instead of *behind* (57)
- Choosing becomes the supreme vocation (45)
- Becoming set free from the past (57)
- Recognition of *failures* as more fruitful than *successes* (46, 48, 49)
- Freeing oneself from thought-patterns of adulthood (57)
- Reviewing life to decipher the enigma of its worth (57)
- Encounter with God (59, 61)

Appendix B

Everybody Dies

by Pat Boone
Everybody dies, everybody dies.
It's a sad but true fact of life.
Everybody dies.
And everybody knows that everybody goes.
You just sing your song, then move along.
Everybody dies.
Wish I didn't have to think about it.
Wish it wasn't so.
But if Einstein couldn't get around it, I'm sure I'll have to go.
So will Mom and Dad and Uncle Harry – baby sister too.
And if death could take the Virgin Mary,
Ya know it won't miss you.
Cause everybody dies, everybody dies.
It's a sad but true fact of life.
Everybody dies.
And everybody knows that everybody goes,
You just get in line until your time.

Everybody dies.
Read it on a bumper sticker,
Pondered it for months.

Born just once, and you'll die twice.
Born twice, you just die once.

Heard folks say they're born again
And they haven't even died.
Others laugh to hide the funerals goin' on inside.
Heard God saw that we're all dyin'
He sent His only Son.
He died, too – but lived again.
Said "I'll show you how it's done."
Think I'd better take the offer – best one that I've found.
Cause after all my slippin', slidin',
Feels like solid ground!
Cause everybody dies, everybody dies.
It's a sad but true fact of life — everybody dies.
And everybody knows that everybody goes.
So meet the Son. He's the One.
Cause everybody dies.

© Spoon Music, Corp 1990

Appendix C

BIBLIOGRAPHY OF REFERENCED BOOKS

The NIV Worship Bible (Grand Rapids, MI: Zondervan, 2000)

James Allen, ***As a Man Thinketh*** (The Savoy Publishing Company, 1903)

Ernest Becker, ***The Denial of Death*** (New York: The Free Press, 1973)

Sam Donaldson, ***Hold On, Mr. President!*** (New York: Random House, 1987)

Will Durant, ***The Reformation*** (New York: Simon and Schuster, 1957)

Marie de Hennezel, ***Intimate Death*** (New York: Alfred A. Knopf, 1997)

Vernon L. Grose, ***MANAGING RISK: Systematic Management of Risk*** (Englewood Cliffs, NJ: Prentice Hall, 1987)

Vernon L. Grose, ***Purpose In A Random World*** (Amazon Digital Services — ASIN B00AB85LT6, 2013)

Vernon L. Grose, ***Science But Not Scientists*** (Bloomington, Ind: AuthorHouse, 2006)

Os Guinness, ***The Dust of Death*** (Wheaton, Ill.: Crossway Books, 1994)

Frederick J. Hacker, ***Crusaders, Criminals, Crazies*** (Toronto: George J. McLeod Limited, 1976)

Mike Huckabee, **Kids Who Kill** (Nashville: Broadman & Holman Publishers, 1998)

Elisabeth Kubler-Ross, **On Death and Dying**, (London: Routledge, Taylor & Francis, 1970)

John Lukacs, **At the End of An Age** (`New Haven & London, 2002)

Sherwin B. Nuland, **How We Die** (New York: Vintage Books, 1995)

Hiroshi Obayashi, **DEATH AND AFTERLIFE** (New York: PRAEGER, 1992)

Marshall B. Rosenberg, **Nonviolent Communication: A Language of Life** (Encinitas, CA, PuddleDancer Press, 2008)

Kathleen Dowling Singh, **The Grace in Dying: A Message of Hope, Comfort, and Spiritual Transformation** (New York: HarperCollins Publishers Inc.)

William Strauss and Neil Howe, **The Fourth Turning: An American Prophecy** (New York: Broadway Books, 1997)

Studs Terkel, **Will the Circle Be Unbroken?** (New York: The Free Press, 2001)

Paul Tournier, **The Seasons of Life** (Richmond, VA: John Knox Press, 1964)

Index

DEATH IS NOT FATAL

A

AAF B-25 medium bomber - 47
AAL Flight 77 - 55
Aaron, Benjamin – 60
ABC This Week – 82
ABC Good Morning, America – 143
ABC Prime Time Live – 82
Abortion – 136
AC&W (aircraft control & warning) – 66
Act of God -- 67, 69
Adam – 187
Aesop – 158
After Life, movie – 60
Agnew, Spiro -- 100
Agnostic – 67
AIDS – 84, 85
Air Force One -- 53
Air Traffic Control – 46
Aircraft crashes -- 30
Alcor Life Extension Foundation (ALEF) – 106
Allen, James -- 170
Allen, Woody – 91
Altruism – 29
American Airlines 757 – 49, 50
American Presidents -- 113
AMTRAK – 140-142
Amy (DeWitt) – 28
Antietam – 53
Apollo – 21, 22, 95, 163
Apollo Command Module – 126
Apollo-Soyuz – 23
Archduke Franz Ferdinand – 61, 128

Arlington National Cemetery – 101, 161
Atheism – 67
Atheist – 67
Atlanta – 50, 80
Atomic bomb – 57
Augustine of Hippo – 122
Auto traffic deaths – 85
Avianca Airlines 767 -- 65, 66
Avianca Flight 52 – 66

B

Baker, James – 53
Bankers Trust – 33
Barksdale AFB, Louisiana – 53
Bastiat, Frederic – 133
Battle of the Bulge – 97
Becker, Ernest – 27, 89
Beijing, China – 101
Belief – its dominance – 139
Better Chevrolet Service – 72
Bible authorship – 112, 113, 115
Biblical options on death – 123
Bigotry – 157
Black September – 16
Blue Angels – 40
Bogata, Columbia – 66
Bohemia – 87

Bombay (Mumbai) India – 193
Boone, Pat – 124, 196, Appendix B
Breaux, Senator – 54
Bryce, James – 189
Burgermeister of Koln – 96
Bush, President George H.W. -- 53
Bush, President George W. – 53

C

Caesar – 167
Cancun – 13
Capitol – 50
Carter, President Jimmy – 110
CBS Newswatch – 143
CBS This Morning – 82
Centrality – 22
CFRB Radio – 82
Chalif, Gail – 80
Charter airlines – 13
Chein, Isador – 99
Cherry pickers – 141, 142
Chessie System – 179-181
Chicago – 50
Chinese Academy of Sciences – 101
Cicero, Marcus Tullius – 72, 95

CIP: Critical Infrastructure Protection – 172-3, 175

Circus risk – 40

Civil War – 53, 100-101

Closure – 68

Cloud computing – 174-5

CNN – 13, 14, 80

CNN Atlanta – 80

Cochran, Jacqueline – 158

Cockpit voice recorder (CVR) – 32, 55

Cole, Thomas – 58

Cologne, Germany – 95

Compes, Peter C. – 96, 97

Compes, Marianna – 96

Congressional Medal of Honor – 159

Conflicting Priorities to Disaster – 145-6

Conover, Katie – 46, 49

Constitution of the United States – 129, 137

Crimes against humanity – 73

Cronkite, Walter – 128

Cryonics – 154

CT (computed tomography) – 89

Cycle of civilizations – 177

Czech government – 87

D

Dallas, Texas – 127

Darwin, Charles – 77

Datacide – 170

de Hennezel, Marie – 84

Death as ultimate risk – 43

Death classic symbols – 144

Death definition – 86

Death One – 187

Death Two – 187

Deathbed – 83

Debi – 13-15, 23, 37, 53-56, 116, 196

Declaration of Independence – 148

Decoration Day – 100

Deferred gratification – 172

DeLeuw, Cather & Company – 178

de Montaigne, Michel – 80

Design Review – 126

Desired Outputs – 24-26

DoD – 91

Donaldson, Sam – 82-83

Doolittle, Jimmy – 159-162

Downey, CA – 126

Dryden, John – 75

Dukaksis, Michael – 100

Durant, Will – 188

Dying twice – 182

E

Earth Day Network (EDN) – 87
Earth Day Shabbat – 88
Earth temperature – 133
Easter – 194
ECMO (Extracorporeal Membrane Oxygenation) – 89
Egyptian pyramids – 100, 104
Eisenhower, President Dwight – 113
Eliot, T. S. – 168
EMP (Electro Magnetic Pulse) – 172, 175
Empire State Building – 47
Enlightenment – 74
Environmentalism – 87
ESP (Extra Sensory Perception) – 164
Evans, Gail – 80
Evil – 73, 74, 77, 138

F

FAA – 91
Frank Lloyd Wright – 100
Fatal and fatalities – 184

Fatality density – 85, 107
Fates: Clotho, Lachesis, and Atropos – 184
FBI – 50
FDA – 91
Fear of uncontrollable – 90
Federal Railroad Administration – 142
Federal Housing Administration – 91
Flight Data Recorder (FDR) – 32, 55
Florida – 109
FOX Green Room – 54-55
FOX News – 46
FOX Morning News – 82
Francis, Robert -- 146
Frank Lloyd Wright homes -- 100
French Revolution – 74
Freud, Sigmund – 59
Fromm, Erich – 59
Frost, Robert – 192
Fyodorov, Nikolai Fyodorovich – 154

G

Gemini, Project – 22
Genesis – 187

George Washington University – 49, 179

German people – 72, 95-96

Germanwings Flight 9525 – 56, 85, 150

Gettysburg, PA – 107

Giant Bible of Mainz – 114

Gingrich, Newt – 53, 54

Global warming – 85, 132

Golgatha – 194

Governor's Select Committee on Law Enforcement Problems – 130

GPS tracking -- 175

Greek concept – 17, 23

Grose, Wesley – 72, 161

Guinness, Os – 149

Gulf War – 53

Gumbel, Bryant – 143

Gutenberg Bible –114

H

Hacker, Frederick J. – 57

Haig, Alexander – 53, 61

Hanford, Washington – 57

Hart, Gary – 100

Hate crime – 134, 135

Havana, Cuba – 109, 111

Hegel, Georg Wilhelm Friedrich – 92

Hegel, Senator – 54

Heinemann, Ed – 158

Herd behavior – 105

HHA – 91

Hitler, Adolf – 72, 73

Hook-and-ladder – 141

Houston, Texas -- 23

Howe, Neil -- 118

Huckabee, Mike – 149

Hudson River – 47

Human death – 27

Human goodness – 76

Human risk – 37, 39

I

I Thessalonians 5:2-3 – 51

Impersonal death -- 173

Information, knowledge, wisdom – 169

Inhofe, Senator – 54

Institute of Nanotechnology – 106

Iran – 193

Israel – 194

J

James, William – 59

Japanese Red Army – 193

Jerusalem of Gold – 84

Jesus Christ – 194

JFK airport – 46, 66, 67

Johnson, Val – 109

Judeo-Christian belief – 148

Jung, Carl Gustav – 59

Jurisprudence of thought life – 135

K

K Street lobbyists – 130

Kelvin, Lord William Thompson – 169

Kennedy, President John F. – 22, 61, 127, 128

Key West, Florida – 109

Kierkegaard, Soren – 59. 105

King, Martin Luther – 119

Kloman, Felix – 33

Known Inputs – 24, 25

Known precedents – 30

Koelnmesse – 95

Korea – 98

Kremlin – 63

Kuala Lumpar, Malaysia – 174

Kubler-Ross, Elizabeth – 16

Kurzweil, Ray – 155

L

Laguardia airport – 46

Landon, Alfred -- 138

Lane, Charles – 138

Lawlessness – 78

LeMay, Curtis E. -- 159-162

Lenin – 63

Lenis, Gustavo – 66

Library of Congress – 114

Life After Death movie – 60

Life defined – 104

Life values – 176

Lightning strike – 67

Lincoln, President Abraham – 61, 107

Lod airport – 193

Long Island, NY – 66

Lord Acton – 131

Los Angeles – 126

Lunden, Joan – 143

Lukas, John – 77

M

MacArthur, General Douglas – 98
Madoff, Bernard – 134
Malaysian Flight 370 – 174
Manhattan Project – 17, 19
Marriage – 136
Marshall Space Flight Center – 21, 165
Maryland State Assembly -- 182
Maslow, Abraham -- 190
McEnroe, John – 66
Mediterranean Sea – 193
Memorial Day – 100
Mengele, Josef -- 147
Mercury, Project – 22
Meserve, Jeanne – 14
METRO, Washington DC –178-183
Mexican President Portillo – 110
Militant Islam – 43
Minneapolis church – 28
Mitchell, Edgar D. – 163, 164
Mitchell, Ollidean – 164
Monterrey, CA – 161
Montezuma's Revenge – 110
Moon – 22, 126-128, 163, 164
Moral death – 121, 124
Moravec, Hans – 155

Morrison, Mark – 106
Moscow, Russia – 63
Moses – 115, 116, 167
MRI (Magnetic Resonance Imaging) – 89
Mumbai, India – 193
Murphy, Michael – 52, 66
Music and death – 103
Muskie, Ed – 100
Muslim – 56

N

Nagasaki, Japan – 57
Nanotechnology – 106
NASA – 20, 21, 95
NASA Safety Advisory Group for Space Flight – 21
National Advisory Committee for Aeronautics (NACA) – 20
National Archives – 39
National Aviation Show – 158, 166
National Gallery of Art – 58
National Transportation Safety Board (NTSB) – 30, 47, 113, 140, 146
NavCanada – 46
Navy Annex – 101
Nazi Germany – 73, 96

NBC The Today Show – 143
Neuroseparation – 106
New York – 46, 54, 140
Newark airport – 46, 47
News media – 140-144
Nicodemus, Waldo – 109
Nietzsche, Friedrich – 105
Noble Savage theory – 74
Nonviolent communication – 73
North American Aviation – 126
North, Oliver – 52
Nothingness – 104
NPR Morning Report – 81
NTSB Go-Team – 140
Nuland, Sherman B. – 84, 149
Nuremberg trials – 73

O

O'Halloren, Chuck – 158
Obayashi, Hiroshi -- 122
Occupational risk – 41
Olson, Barbara – 52
Olympics – 16
Order of two deaths – 125
Organ transplantation – 86
Osama bin Ladin – 48
OSHA – 91

Ovid – 109, 178

P

Pain – 168, 171
Pakistan – 193
Pascal, Blaise – 139, 176
Paul, apostle – 88
Pearl Harbor – 53
Pentagon – 49-51
Pepto-Bismal – 110
PET (Positron Emission Topography) – 89
Phyllis Grose – 29, 49-51, 81
Physical death – 121, 125
Plato – 59, 167
Poincare, Henri -- 65
Political apocalypses – 132
Political correctness – 134, 139, 170
Political half-life – 100
Political impotence -- 137
Political involvement in death – 135-137
Political manipulation – 90
Polling – 138
Pomerantz, Steve – 50
Pope Francis – 132
Port Authority of New York – 49

Post-human existence – 151
Post-mortem photography -- 63
Power of belief – 139
Presidential oath and the Bible – 113, 114
Princess Diana – 144
Prisco, Guilio – 154
Psychedelics – 135
Pyramids of Egypt – 100

Q

Quayle, Dan – 100
Questioning death – 111

R

Rabelais, Francois – 92
Raddatz, Martha – 81
Rage – 68
Reagan, Governor Ronald -- 130
Reagan, President Ronald – 61, 161
Reingold, Les – 140, 142
Religious Action Center of Reform Judaism (RAC) – 88
Renaissance Man – 19
Resurrection – 56, 116-119
Resurrection of Jesus Christ – 194, 195
Rheingold, Howard L. – 59
Rhine River – 95, 96
Rice University – 128
Risk as a flower – 37, 38
Risk assessment – 44
Risk definition – 34
Risk dimensions – 37
Risk exposure -- 40
Risk ignorance -- 38
Risk management principles – 45
Risk specialties – 36
Risk Totem Pole – 181
Risk-loss linkage – 35
Risk-taking – 39
Road One – 192
Road Two – 192
Roberts, Cokie – 82
Rockefeller, Nelson – 100
Rogers, Carl – 190
Roosevelt, President Franklin D. – 138
Rosenberg, Marshall B. -- 73
Rousseau, Jean Jacques – 74

S

Sacred Way or Divine Road – 101, 102
Saddam Hussein – 53
San Fernando Valley, CA – 126
San Francisco de Paula, Cuba – 109
San Luis Obispo, CA – 31
Santayana, George – 91
Sapir, Naomi – 84
Saving lives fallacy – 92
Schere, Mike – 72
Schizophrenia – 97, 98
Science of evil – 90
Scientism – 99, 188
Scott, Jon – 47-49
Scottsdale, Arizona – 106
Seattle, Washington – 109
Secular Humanism -- 104, 148-9
Self-actualization – 190
Self-centeredness – 172
Self-identity – 190, 191
Self-managed death – 185
Shakespeare – 64, 167
Shaw, George Bernard – 140
Shepard, Alan – 163, 164
Technological Singularity – 151
Silos – 18
Singh. Kathleen Dowling – 95
Sinisterra, Captain Carlos – 66

Sky News Arabia – 174
SMART (Systems Methodology Applied to Risk Termination) – 15
Silver Snoopy medal – 22, 23
Social media – 174
Socrates – 46, 59, 126
Soviet occupation of Bohemia – 87
Space Shuttle Challenger – 144
Space Shuttle Columbia –165
Space Shuttle Discovery – 166
Spokane Daily Chronicle – 57
Spokane, Washington – 57, 72
SS "City of Havana" – 109
Stafford, BGeneral Tom – 22
Stem cell research – 147, 188
Stone animal statues – 101
Stovepipes – 18
Strauss, Richard – 64
Strauss, William – 118
Strum, Marv – 109
Suffolk County Medical Examiner – 83
System – 24
Systems approach – 17
Syria – 193

T

Index

Tahlequah, Oklahoma – 164
Taj Mahal – 100, 104
Technology and death relationship – 88
Tehran, Iran – 193
Tel Aviv, Israel – 193
Tennyson, Alfred Lord – 28, 58
Terkel, Studs – 98
Terror of death – 57
Terrorism – 57
Thanatology – 16
The Conquest of Granada – 75
The Literary Digest – 138
The Road Not Taken – 192
The Washington Post – 183
Thirteen Tombs of the Ming Dynasty – 101
Three Mile Island – 16
Thunderbirds – 40
Time magazine – 82, 152
Tipler, Frank – 154
Thomas Jefferson – 189
Total Person – 119, 120
Tournier, Paul – 118, Appendix A
Transhumanism – 151
Transplanting organs – 147
Truman, President Harry – 98
TWA Flight 800 – 80, 81, 156
Two deaths – 121

Two times of death – 124

U

US Marine Corps – 101
Uniform Anatomical Gift Act (UAGA) -- 86
United Airlines Flight 93 – 54
Unintended consequences – 137
United Nations – 171
USAF C-5 Galaxy – 193

V

Valujet Flight 592 – 13
Vengence – 68
Vester, Linda – 55
von Braun, Wernher – 21
Voyage of life – 58

W

Wallace, Chris – 80
Wannabee syndrome – 105
War – 78, 97, 98
War crimes – 73
Washington Court hotel – 80

Washington Metropolitan Area Transit Authority (WMATA) – 16, 178-182

Washington Post editorial – 183

Washington, President George – 173

Weltanschauung revision – 148, 153

White House – 195

Will Rogers – 118, 195

Williams, Ted – 106

Wisconsin – 28

WMD (weapon of mass destruction) – 57

World Trade Center – 46, 49, 51

World War I – 61, 73, 78, 128

World War II – 17, 18, 78, 100

World War III – 51

Worldwide environmental network – 87

Worship of the earth – 87

Y

Yamasaki, Minoru – 51

Yom Kipper War – 193

Z

Zilboorg, Gregory -- 59

Author Profile

Described in *Business Week* as a founding father of the application of systems methodology to managing risk, he enjoys worldwide recognition as an authority in that field. A physicist by education, he has served as an executive in three major corporations, university professor in Europe as well as the United States, and consultant to such firms as AT&T, EXXON, and IBM.

He originated the widely-adopted **SMART** (*Systems Methodology Applied to Risk Termination*) technique for managing every type of risk – legal, political, social, economic, and technological – which was successfully utilized to combat terrorism at the 1984 Olympics in Los Angeles.

The Peoples Republic of China invited him in 1981 to address the Academy of Sciences in Beijing on the systematic management of risk.

President Reagan appointed him to the National Transportation Safety Board in 1983 and the National Highway Safety Advisory Commission in 1986. The White House assigned him for one year as Expert Consultant to the NASA Chief Engineer where he pioneered a model for commercial enterprise in space. In 1997, Vice President Gore solicited his expertise for the *White House Commission on Aviation Safety and Security*.

His best-selling book, ***MANAGING RISK: Systematic Loss Prevention for Executives*** was called "the most influential book on the subject in this decade." He has written two other books: ***SCIENCE BUT NOT SCIENTISTS: How Everything Began – Chance or Creation?*** and

PURPOSE IN A RANDOM WORLD: Have you ever wondered WHY?
His professional papers have been published internationally in over 60 journals and periodicals.

A featured guest on the *Today Show, Good Morning America, Prime Time Live, CBS Newswatch, ABC 20/20, BBC-London, O'Reilly Factor* and many other television programs, he has given over 100 interviews on CNN as their Risk and Aviation Analyst. He gave over 170 interviews on the 1996 explosion of TWA Flight 800. Dr. Grose is a FOX News Contributor and was being interviewed just as UAL 175 impacted the World Trade Center Tower 2 on 11 September 2001. His viewpoints have been published in such periodicals as *Time, USA Today, US News & World Report, Chicago Tribune, Los Angeles Times, Washington Post,* and *Christian Science Monitor.* He was honored with Whitworth University's 2013 Distinguished Alumni Award.

His biography appears in ***Liftoff*** by James C. Hefley that describes the personal faith of astronauts and space scientists. He is listed in *Dictionary of International Biography, Men of Achievement 1973, International WHO'S WHO of Intellectuals,* and *WHO'S WHO In The World.*

Dr. Grose is married to Phyllis Jean (nee Heine). They have six children, 28 grandchildren, and 5 great-granddaughters.

Contact Information:
Dr. Grose can be contacted for related interviews, counsel, or correspondence at vgrose@omegainc.com

CPSIA information can be obtained
at www.ICGtesting.com
Printed in the USA
LVOW02s1723260916
506165LV00028B/34/P